TAXONOMY AND GENETICS OF OENOTHERA

MONOGRAPHIAE BIOLOGICAE

EDITORES

F. S. BODENHEIMER

Jerusalem

W. WEISBACH

Den Haag

VOL. VII

Springer-Science+Business Media, B.V

TAXONOMY AND GENETICS OF OENOTHERA

Forty years study in the cytology and evolution of the Onagraceae

BY

R. RUGGLES GATES
(Cambridge, Mass.)

Springer-Science+Business Media, B.V

ISBN 978-94-011-7943-0   ISBN 978-94-011-7941-6 (eBook)
DOI 10.1007/978-94-011-7941-6

Softcover reprint of the hardcover 1st edition 1958

CONTENTS

FOREWORD

At the beginning of this century the mutation theory of evolution, based by HUGO DE VRIES on many years of culture experiments mainly with *Oenothera Lamarckiana*, took the world by storm when published in 1901. As early as 1889 DE VRIES had shown that various plant "monstrosities" are inherited. His breeding experiments with other plants involved the rediscovery of MENDEL's laws and the resuscitation of MENDEL's breeding methods, which had been in abeyance in the long period from 1865 to 1900. MENDEL began his crossing experiments in 1854, DE VRIES began his Oenothera cultures in 1886. The genus Oenothera was thus the first after Pisum to be the subject of extensive genetic experiments. Thus was introduced the method now so widely pursued. It may be remembered also that DE VRIES drew the distinction between fluctuations and mutations, but he regarded all fluctuations or continuous variations as non-heritable. Both can be equated with the continuous and discontinuous variations of BATESON.

DE VRIES' conception of mutations corresponds with the "single variations" of DARWIN, except that he regarded each mutation as producing a new elementary species, now generally called microspecies. He said in the introduction to his Mutationstheorie, that the characters of organisms are built up from units which are as sharply distinguished from each other as the molecules of chemistry. These units can be united into groups, and in related species of plants and animals the same units and groups are repeated. This is still orthodox genetics, despite all we know now of genes and their relation to ontogeny; but in one aspect DE VRIES was misled. In Oenothera, the catenation of the chromosomes, discovered long afterwards, leads to the formation of the complexes of RENNER, each of which is a large aggregate of characters based on many genes. DE VRIES naturally assumed that these adhering complexes of characters were single mutations in origin.

The observation that mutations which were later discovered to be trisomics, triploids or tetraploids also showed differences in all parts of the organism tended to confirm this point of view, but in these cases again whole chromosomes with all their contained genes were involved. These cases are necessarily retained as a part of the conception of mutation in organisms, and are now known to have been significant in the formation of many species and even genera of plants, although in animal species the processes of chromosome evolution have generally taken other directions.

As already mentioned, DE VRIES regarded all fluctuations or quantitative variations as non-inherited. Now we know, from the work of MATHER and many others, that many quantitative differences are inherited, and frequently on a multiple basis. In the work here recorded, it is clear, for example, that small differences in petal size between strains of an Oenothera species are inherited; and in inbred lines, the difference in mean petal-length may be constant even to fractions of a millimeter. Thus each species of plants may have characteristic quantitative as well as qualitative characters, but the former are more difficult to recognize because of the frequently large element of environmentally induced fluctuations.

The biometric and the Mendelian schools were built upon the study respectively of quantitative and qualitative characters, each being unable to see any merit in the other's observations. Sir RONALD FISHER[1] was largely responsible for reconciling the Mendelian and the biometric points of view. Now it is clear that some of the biometric genes, for example in dwarf mutations, are in effect qualitative, while many qualitative characters such as skin colour in man can be treated metrically.

The work of MORGAN and his school with Drosophila emphasized that many gene mutations are small, although a few show great changes from the type. The vast development of Drosophila genetics more recently under the leadershsip of MULLER and many others has led to a greater knowledge of this fly than of any other organism except man himself.

In the half century since the work of DE VRIES inaugurated the new era of plant breeding there has been a tendency, especially among zoologists, to narrow the conception of mutations until, in many minds, they are confined to what are now known as gene mutations or single locus substitutions. After several different kinds of mutations, such as trisomics, tetraploids and chromosome translocations, had been analyzed and shown to depend on quite different kinds of germinal (nuclear) changes, it was proposed (GATES 1931) to classify mutations on a germinal basis. "The term mutation should be used in the generic sense, including various classes or categories of mutational change". To confine the term to gene mutations is unduly restrictive in eliminating the various other kinds of changes which contribute to evolutionary processes.

The term "catenation" was suggested for the process in which non-homologous chromosomes exchange ends, thus producing permanent rings or chains in many species and greatly altering their genetic behaviour. It is now known to occur in many other plant genera, but usually without the presence of balanced lethals.

My own work with Oenothera, a brief resumé of some features of which is included here, began in 1907. In that year the red-budded

---

[1] "The correlations between relatives on the supposition of Mendelian inheritance." *Trans. Roy. Soc. Edinb.* 52: *399-433.* 1918.

dominant gene mutation *rubricalyx* appeared in my cultures of
O. rubrinervis (See GATES 1911). The same mutation has since
appeared in at least two widely different species, (a) in *O. perangusta*,
a small-flowered species north of Lake Superior, at a single wild
locality (GATES 1950, 1958); (b) in the South American species,
O. affinis (subgenus Raimannia), found in Argentina and Uruguay
(TANDON & HECHT 1953). The conception of parallel mutations was
based (GATES 1912) however on the occurrence of a trisomic *lata*
mutation in *O. biennis* exactly similar (except in the small flowers)
to the *lata* mutation from *O. Lamarckiana*.

Later cytological studies (GATES & GOODWIN 1931) showed that
the ring – pairs of chromosomes in meiosis are frequently interlocked
because the free ends pair specifically and simultaneously while the
middle portions spread apart under a mutually repelling action.
This is partly related to the median position of the centromere, and
in any case contrasts greatly with the condition in many plant genera, in which the chromosomes pair in meiosis and twist around each
other throughout their length.

The bulk of this monograph is devoted to an intensive taxonomic
study of the genus Oenothera in Eastern North America. The
experimental work in growing the plants from hundreds of localities in Canada and the United States each year was done in the
Courtauld Research Laboratory in Regent's Park, London, endowed
by Sir William COURTAULD in 1933. The results of three years'
research were published in *Phil. Trans. Roy. Soc.* 1936. This was
followed by four years more of cultures from fresh localities, confirming and extending the previous results. Greater familiarity with
all the forms made it possible to reduce several of the "species" to
varieties of species already described. The whole work is thus an
example of taxonomic genetics, on a large scale and of the most
intensive character. The results give a picture of the geographic
distribution and relationships of these forms. Various problems of
taxonomic description such as the relation between qualitative
and quantitative (measureable) characters in the delimitation
of species, are considered.

In addition, the main contributions of RENNER, CATCHESIDE,
DAVIS, CLELAND and others have been added to give a full picture
of the forces and processes which have conspired together to produce
the genus and family as we now know it. Much more remains to be
done, particularly in determining the chromosome-end relationships
among all these new forms in the eastern half of North America, but
the methods of evolution within the group are now clearly outlined
by the discoveries of the last fifty years. It is clear that taxonomic
genetics and cytological genetics are complementary methods in
working out in greater detail the paths of evolution within the
genus.

Some might be inclined to treat all these new forms as microspecies, but many of them are quite as distinct from each other as *O. biennis*, *O. muricata* and *O. parviflora*, described in the 18th century and, they in no sense fit into the limitations of those species.

Finally, the long delay in publication has resulted, partly from the war and partly from the fact that I have since been deeply engaged in human genetics and anthropology. But the main results of others have been included to bring the work in the field here considered up to date.

Those who are not geneticists should read first the Summary and Conclusions.

Cambridge, Mass. January 1958.

EVOLUTION

Conditions of Evolution in Oenothera

Notwithstanding the immense amount of breeding and cytological work done with Oenothera, the taxonomic treatment has been in an unsatisfactory condition. Professional taxonomists have largely avoided consideration of the modern work, probably because many of the critical characters are not well preserved or are difficult to observe in herbarium specimens. The subgenus Oenothera (formerly Onagra), in which major interest has centred since the experiments of DE VRIES, shows a wide range of variation. It extends over most of the north American continent and shows many significant characters which have presumably originated through mutation. These variations affect every part of the plant. They include many characters of leaf shape, size and colour; of the stem (erect, oblique, bent at the tip, or crozier-shaped), flowers, and pubescence; as well as branching, comparative time of flowering under uniform conditions, and other reactions such as a weaker or stronger tendency to the biennial habit or the development of anthocyanin in certain organs such as the leaf midribs or marginal glands, and in the papillae on the stems and buds. These papillae not only vary greatly in size in different strains but may be red, pink, or colourless on the stem, ovary, sepals or sepal tips. GERSHER (1921) showed that in *O. muricata*, *O. ammophila* and *O. cruciata* the (oblique) stem continues bending in darkness. The bent tip, however, also shows photonastic movements. The maximum daily movement of the stem tip takes place at about the time of flowering, such movement ceasing at the end of flowering. Different species vary much in the degree and time of bending of the stem tip. In *O. ammophila* the tip takes a crozier shape (due to the *percurvans* complex), in *O. muricata* it is less marked *(curvans)*, and in many species it is only slight or temporary, disappearing early in the flowering period. It would be interesting to know if auxin distribution is involved in these cases.

When I began a genetic survey of the Oenotherae in Eastern Canada in 1932, it seemed possible that they could show such uniformity that there would be little in the way of variability to record. On the contrary, a most surprising range of variations was disclosed. After three years of cultures, which showed that each seed collection was generally uniform and bred true except for an occasional trisomic or tetraploid mutation, a monograph (GATES, 1936) was published which included 36 new names (species and varieties) in Eastern Canada and adjacent States.

On the basis of many years of cytogenetic research with Oenothera, conclusions were reached (GATES, 1933) regarding the origin, phylogeny and geographic dispersion of the genus following the last glaciation, and most of the species then known were briefly characterized. They also received a preliminary taxonomic consideration at an earlier period (GATES 1915). It was concluded that the primitive southern forms had large flowers (petals 40 mm. or more in length) and that as they moved northward a series of dominant mutations gave rise to progressively smaller flowers (petals 10 mm. or even less). That this happened independently in different lines of descent is indicated by the fact that some of the small-flowered forms in British Columbia resemble the Californian *O. Hookeri* and are probably descendants from it. The northward migration up the Pacific coast has been contemporaneous, as elsewhere, with the development of small, self-pollinating flowers. The small-flowered form from British Columbia was formerly (GATES, 1915, p. 29) treated as *O. Hookeri* var. *parviflora*, [1] based on a specimen in the British Museum from Kamloops. B.C., collected by JOHN MACOUN in 1889. This sheet bore the name *O. biennis* var. *hirsutissima* GRAY. In this specimen the petals were only 14 mm. long. Another specimen (1. c., p. 30) in Herb. Lindley, collected by DOUGLAS in "N.W. America", 1827 (?) had somewhat larger flowers (bud cone 25 mm.) and leaves with red midribs. A third specimen differs from the last only in bud cone 11 mm. nearly free from pubescence. No doubt further investigations will disclose many minor small-flowered types in that region, partly derivatives of *O. Hookeri*.

The change in flower size in Oenothera was accompanied by two selective advantages, (1) the small flowers are self-pollinated, not requiring insect (mainly moth) visits, so that every flower produces a large number (ca. 300) seeds, (2) in the smaller flowers there is a great saving in material and especially in pollen production, since the anthers are directly in contact with the stigma in the bud. The anthers are also very much smaller, and pollen production is greatly reduced, as in the evolution of the Compositae. This mechanism for self-pollination does not, however, wholly exclude crossing, which occurs in the wild with a frequency which is difficult to estimate (see GATES 1958), but is probably very low. As the pollen is too sticky to be carried by the wind (except by gusts), such crossing as occurs between different colonies is probably the result of insect visits in the evening when the flowers first open. Even then, self-pollination will have occurred previously.

These conditions are associated in Oenothera with catenation of the chromosomes. As the large-flowered species such as *O. Hookeri*

---

[1] As *O. Hookeri* var. *parviflora* was a nomen nudum, it may be briefly characterized as follows: A specie differt, flores minores, petala 14 mm, citing the MACOUN sheet as the type.

and *O. argillicola* have seven free pairs or a ring of 4, while nearly all the small-flowered forms have a ring of 14, it is evident that increase in chromosome catenation up to complete linkage developed rapidly and concomitantly with the small flowers. That new species have arisen in Oenothera by crossing of forms with catenation was indicated by GATES (1928). As catenation, combined with the presence of balanced lethals, prevents the segregation of homozygous types in the offspring, this condition results in species which breed true although highly heterozygous (composed of two complexes, one frequently functioning only, or chiefly, in the pollen and the other mainly in the egg cells). Any advantages derived from hybrid vigour are thus secured, and continuous self-pollination does not result in increased homozygosity. The advantages of self-pollination in increased seed production were recognized (GATES, 1915, p. 38) long before it was known that these forms are crypthybrids owing to chromosome catenation.

We know that catenation has developed independently in Gaura (BHADURI, 1941), Clarkia (HÅKANSSON, 1931), Godetia (HÅKANSSON, 1925, 1941) and Eucharidium (SCHWEMMLE, 1926) of the Onagraceae and in many other genera of Angiosperms (GATES, 1951). It is thus a condition which has evolved, like apomicty, independently many times, but in most genera outside the Onagraceae it is apparently not accompanied by balanced lethals. The advantages in Oenothera of catenation associated with balanced lethals are clear, but its possible relation to phylogeny in genera outside the Onagraceae is not known.

In the offspring of *gaudens* [h.]* *franciscana*, which is the hybrid *O. Lamarckiana* X *O. franciscana*, DAVIS (1947) found an etiolated class of seedlings which corresponded to [h.] *franciscana* [h.] *franciscana* in the cytoplasm of *O. Lamarckiana*. In the F_6 of this culture, seeds which had been subjected to high temperature (90° C.) for five days produced ten plants, one of which gave seeds with a germination of only 36.2%. In the F_7–F_9 from this plant the etiolated (sublethal) seedlings were replaced, through mutation, by a zygotic type which was completely lethal, i.e. the embryo failed to develop. A sublethal has thus mutated to a lethal condition. A sublethal type of etiolated seedlings is known to occur in many interspecific Oenothera hybrids, but how the balanced lethal condition has generally arisen remains unknown. It probably occurred as a natural gene mutation from the sublethal to the lethal condition.

The general conclusions regarding the phylogenetic history of Oenothera (GATES 1933) have since been confirmed by CLELAND (1940, 1949), who repeats many of my original conclusions regarding the factors of evolution in the genus. He also (1944) reports catenation in the onagraceous genera or subgenera Hartmannia, Lavauxia and Raimannia, with balanced lethals in the last genus. The South American *O. campylocalyx* and certain related species are also found to have balanced lethals and chromosome catenation. The subgenus Raimannia appears to be largely developed in temperate South America, and although *O. rhombipetala* in Minnesota may have seven free pairs, in other localities it has ⑭* or other catenations (HECHT 1950). SCHWEMMLE and his associates have shown that in *O. Berteriana* SPACH (Chili), *O. odorata* JACQ. (Chili) and *O. mollissima* L. (Argentina) chromosome rings are present. SCHWEMMLE (1928a) found three types in the cross *O. Berteriana* × *O. odorata*. Type III selfed gives an F_2 which is self-sterile (homozygous for a pollen sterility factor). On one branch of an F_2 plant the chromosomes doubled. From such a branch a *gigas* race could be obtained. In a further analysis (SCHWEMMLE, 1928b) *O. Berteriana* & *O. odorata* are both shown to be isogamous heterozygotes. ARNOLD (1955) gives a genetical analysis of five trisomic mutations of *O. Berteriana*, and ARNOLD & BINA (1957) have recently described three trisomic mutations in *O. odorata*.

SCHWEMMLE (1927) had previously made the cross *O. Berteriana* X *O. muricata*. *O. Berteriana* is annual with a loose rosette, and smaller chromosomes which could be recognized in the F_1 hybrid. Short chains of chromosomes could be seen in the F_1 at diakinesis. Thus some hybrids between these two subgenera succeed, even between species whose homes are in the north and south temperate zones, respectively. It appears that all the genetic peculiarities of the present Oenotheras in North America have developed independently in South America. OELKRUG (1934) studied *O. Berteriana* X *O. mollissima*. The reciprocals gave the same catenation. From reciprocal crosses between South American Raimannias and *O. Hookeri*, KISTER (1955) finds that death of the embryos results from disharmonic exchanges between the hybrid nuclei and the proplastids of *Hookeri*. HARTE & BISSINGER (1952) studied the pollen sterility in hybrids of *O. Hookeri* and concluded that two factors, *fr* and *ster*, affect all stages of anther development. While *fr* prevents the accumulation of reserves, both factors cause a failure of secretion from the tapetum to the pollen mother cells.

O. argentinea has been shown (SCHWEMMLE & ZINTL, 1939) to be homozygous, with all good pollen but much smaller pollen grains than *O. Berteriana* and *O. odorata*. Crosses of these two species with *O. argentinea* give twin hybrids and also plasmatic effects. HAUSTEIN (1952) introduced the system of numbering the chromosome ends in species of Raimannia.

HECHT (1950) has studied the genetics of 14 species of Raimannia mainly from South America. The catenation ranges from 7_{11} to ⑭.

---

* See note on page 101.

Strains of *O. Drummondii* from Texas and from Lower California had generally 7 free pairs, yet when strains were crossed the hybrid showed ⑫, indicating a long separation during which interchange of chromosome ends took place. As in Oenothera, the species with large flowers generally have free pairs of chromosomes or small rings, indicating their more primitive character. TANDON & HECHT (1953) have investigated six races of O. *affinis* (= *O. berteriana*) from Argentina and Uruguay, three of them having ⑭ and three having 7_{11}. The strains differ in such features as petal-size (petals ranging from 22–42 mm. in length), and length of hypanthium (65–115 mm.). Colour of hypanthium is controlled by a single dominant gene, RR and Rr being red, rr pink-green. The gene for red hypanthium has thus arisen independently at least three times: (1) from the *O. Lamarckiana* complex (GATES 1911), (2) from *O. perangusta* north of Lake Superior (GATES 1950, 1958), (3) in the South American Raimannias. These Raimannia races show a maximum of three changes of chromosome ends.

O. *rhombipetata* NUTT. from Oklahoma shows the beginning of catenation but lacks balanced lethals and has developed instead a self-sterility mechanism. A strain of this species from Michigan is, on the contrary, self-fertile but has ⑭ and probably has balanced lethals. The hybrid between them, with ⑩, showed that these 10 chromosomes differ in their end arrangements in the two strains. Crosses between members of the two subgenera failed in all cases except in *O. parodiana* MUNZ from Buenos Aires X *O. Hookeri*. This hybrid was extremely vigorous but sterile, showing the sporadic way in which inter-sterility developes in evolution[2]. On the other hand, such pairs of species as *Drummondii* and *laciniata* when crossed produce a hybrid with seven free pairs or ④ .It is thus clear that the differentiation of species through mutations is quite independent of the cytological rearrangement of chromosome ends.

HAGEN (1950) has investigated the cytogenetics of *O. campylocalyx* and two other South American species of Oenothera, from seeds collected in Argentina, Peru, Bolivia and Chile. He confirms SCHWEMMLE (1927) that the Raimannias and the South American species of Oenothera have smaller chromosomes than the North American species of the latter subgenus. Strains of all the species investigated have one complex in common, the other being unrelated and the usual interchanges of ends have been taking place. These South American forms are phenotypically intermediate between the North American Oenotheras and the Raimannias. They thus form a link with the North American forms but are nearer the Raimannias, from which they may have been derived.

The catenations in a large number of wild North American species

---

[2] WEIDNER-RAUH (1939) has made a study of partial sterility in several South American species.

and forms of Oenothera have been determined, and also in many hybrids (GATES & FORD 1938, JACOB 1940, SIKKA 1940, PATHAK 1940, CLELAND 1949). See Table IV, p. 69. On the assumption that like ends of chromosomes alone pair, indications of relationships can be obtained in the hybrids and in wild strains, and the sequence of translocations between non-homologous chromosomes may to some extent be determined. Chromosome ends may thus, within limits, be homologized by their pairing or non-pairing behaviour. This will be referred to again later. The fact that each species or microspecies is composed of two complexes, one of which (or some features of each) is more or less suppressed by phenomena of dominance, adds to the difficulties in drawing conclusions from a comparison of phenotypes. The characters of the hidden complex are only brought to light by out-crossing.

CLELAND (1944) concludes that "the term 'species' should not be used in connection with this assemblage at all". That, however, is a counsel of despair which is unacceptable to taxonomists, whose job it is to consider and classify the phenotypes of organisms. All genetical, cytological, ecological, distributional and paleontological evidence which can be brought to bear to aid in explaining how the many differences have arisen — everything having a bearing on the evolution of the genus — must be brought into play in an understanding of what has been happening. But the taxonomist's primary job is to classify hereditary phenotypes, describe them and apply names to them, so that others can identify the same or similar forms from his description. The geneticist may then speculate regarding their relationships and past history, using all the evidence available from any source. That, however, is secondary and is only possible after the taxonomist's first function has been performed. The geneticist as such may have more to say about phylogeny, but he cannot usurp the function of the taxonomist and describe as "new species" two forms which are admitted to have practically no phenotypic difference, simply because they do not intercross[3].

It appears that the term "microspecies" will have to be used in entomology for these closely related "genetic species", as it is in botany. Intersterility is a useful criterion of species when used in connection with significant phenotypic differences, but its use (with failure to mate in animals) as the criterion of species could only lead to taxonomic chaos. The absence of intersterility does not mean that the two interfertile forms necessarily belong to one species.

In Oenothera the condition of specific interfertility holds, for practically all the forms can be freely intercrossed, and yet there is a huge amount of inherited phenotypic difference to be considered

---

[3] The subject of species in relation to sterility has been discussed elsewhere (GATES, 1948 Chapter XII and GATES 1951).

and subjected to classification. But Oenothera is to some extent a law unto itself. How far its problems, taxonomically considered, are markedly different from those of other polymorphic genera, such as *Rubus* or *Crataegus*, is difficult to say. Species such as *O. biennis* L., *O. grandiflora* SOLAND., *O. parviflora* L. or *O. muricata* L., although so long recognized as specific entities, are no more distinctive in their phenotype than many species, such as *O. argillicola* MACK., *O. angustissima* GATES or *O. perangusta* GATES, more recently described. Indeed, the differences between *O. angustissima* and *O. perangusta* are more marked than between *O. biennis* & *O. muricata*, neither of which has been found wild in America. Similarly, *O. Hazelae* and *O. grandifolia*, although both came from Nova Scotia, are more diverse than many of the earlier described species. Their characters have not been observed because by common consent professional taxonomists have left the genus alone.

A word may be added here regarding the relation of taxonomic to genetic description. In a polymorphic genus like Oenothera, which has many well marked species as well as many microspecies, the question of the relationship between taxonomic and genetic description becomes important. It appears to me that the lumpers and the splitters among taxonomists have equally important but complementary roles to play. The former point out the resemblances between the various forms in a group, submerging the minor differences. The latter, which inevitably include the geneticists, emphasize the differences and seek to find a place, even for the minor ones, in their classifications.

Having spent considerable time in study of the large collections of Oenothera in the Gray Herbarium of Harvard University, I sympathize with the taxonomists who lump together all forms which can be seen to be similar in herbarium sheets. At the same time I recognize that some of the geneticist's microspecies, whose distinctness is determined by breeding experiments, are clearly distinguishable by certain characters in herbarium sheets; others lose their distinctive characters in pressing and drying, although such characters are discernable by those who have seen the plants in living culture. If the two sets of species were mutually exclusive, or if the microspecies fitted into the categories of the previously described species, the problem could be solved by having two sets of classification, one for ordinary or Linnean species and one for microspecies (including Jordanons and biotypes). But such is by no means the case, and this prevents any simple solution. The buds of *O. ammophiloides* var. *laurensis*, for instance, are easily recognizable with a hand lens on account of the many long hairs arising from red papillae on the sepals; and the bud characters of *O. Victorini* are also distinctive in an herbarium sheet, at least to one who has seen them in cultures. In the absence of any sharp demarkation between these two kinds of species — there is perfect gradation from "species" to "microspecies" — one must rely upon descriptions rather than specimens of many of the latter.

In general, the ordinary taxonomist will continue to rely upon herbarium specimens, and the geneticist upon observation of living plants. These two methods are not in conflict. Rather, they supplement each other, the study of microspecies affording a further analysis of species. With fuller analysis it may become possible to group many of the microspecies under particular Linnean species, but such a grouping is likely to be to a considerable extent artificial. The terms "Linneon", "Jordanon" and "syngameon" are often used in these attempts (GATES 1938), but in the subgenus Oenothera the term syngameon has no meaning, as all the species are interfertile. The term "species" has been used throughout this paper for brevity, but many of the forms considered here, including most of the "varieties", can best be regarded as microspecies. Some taxonomists will prefer to call many of them subspecies.

There is also the fact that among the small-flowered group with complete catenation a hybrid between any two forms will generally breed true and, if sufficiently distinctive, constitute in all essentials a new species, subspecies, microspecies or variety. The concept of species or variety must, therefore, differ in certain respects from these terms as applied in more orthodox genera. But this does not absolve us from describing, in so far as may be, the most conspicuous of the variant types involved. This great amount of diversification has taken place in the post-glacial period. A southern genus has moved northward and undergone what can only be regarded as an extremely rapid evolution.

When we realize the considerable number of Oenothera species which have evolved, even on a conservative definition of species, in the post-glacial period of ten to twenty thousand years, and compare this with the estimate of 50.000 to 1.000.000 years required for the development of a species of mammal, we understand how little there is in common between the conceptions of species in different organic groups.

One must suppose that the genus Oenothera originated either in Central or South America. In the absence of fossil evidence, one must rely mainly on distribution, phenotypic characters and chromosome-ends in reaching general conclusions. BOEDIJN (1924) assumed a South American origin for the subgenus Raimannia, formerly known as Euoenothera, which included such species as *O. mollissima*, *O. longiflora* and *O. odorata*. These are annuals, and *O. grandiflora* (in the former subgenus Onagra, now Oenothera) has a strong tendency in the same direction. The present subgenus Oenothera might easily have been derived from South American species of the subgenus Raimannia, the pollen structure being the same. W. H. HUDSON (1923) found a very fragrant, large-flowered species abundant

on the pampas near the rivers LaPlata and Salado. BOEDIJN believed his accompanying figure to represent a large-flowered species
with the habit of *O. grandiflora*, but from HUDSON's account of its
slender, grass-like stems (with flowers an inch in diameter) it must
also belong to the subgenus Raimannia.

The paleontological record and other evidence makes it clear that
rates of evolution in organisms vary enormously. The genus is
spreading northwards in Western Canada, following man's activi-
ties in removing forests and disturbing the soil. Oenotheras flourish
best on sandy soil disturbed by human activity. Railway banks,
roadsides and sandy beaches are favorite haunts, but they can also
succeed on clay soils or even spread over fully grassed slopes if
conditions are exceptionally favourable.

There is every indication that this group of Oenotheras is now
rapidly evolving in the areas of its multiformity. This includes many
parts of Ontario, Quebec, New Brunswick and Nova Scotia. On the
other hand, the many forms cultured from the Canadian prairies
show little variability and have all been included in one species,
O. insignis BARTL. As will be seen, this species is in process of being
adapted to prairie conditions by shortening the stems and producing
fruits at the ground level, but this adaptation is lost in plants which
develop as biennials in England, only showing the habit in annual
plants grown directly from prairie seeds. This adaptation in ecological habit has already taken place in such genera of the Western
States as Lavauxia and Pachylophis, considered by some as subgenera of Oenothera (MUNZ 1931).

As regards processes of climatic adaptation in Oenothera, HUNGER
(1913) found that *O. Lamarckiana* seeds germinated very quickly
(three days instead of a week or more) in the tropics at Buitenzorg,
Java, and produced the usual mutations (*ca.* 8%). But the plants
remained rosettes and formed no stem even after six months of
luxuriant growth. Similarly, when plants from the Lancashire coast
were grown in a tropical greenhouse (GATES 1912b), those resembling *O. grandiflora* finally formed a stem and flowered after remaining rosettes for ten months. But 28 plants belonging to the more
strongly biennial *O. Lamarckiana* remained rosettes in continuous
growth for nearly two years. In this way a stem without internodes
and covered with leaf bases was gradually formed, reaching over
6 inches in height and resembling in habit a small Cycad. The Cycad
type of stem may thus have arisen as a response to the tropical
conditions where most of them reside.

Why Oenothera should be so relatively uniform over this great
area in Western Canada is not clear, unless it be that the single form
able to survive in the conditions has no others with which it can
cross, but there appears also to have been a dearth of mutations.
Certain other species are known to occur, e.g. at Winnipeg (GATES

1915, p. 24), and without being an extreme splitter one might distinguish the Canadian prairie form from *O. insignis* BARTL. at Duluth, Minnesota.

Crossing will not, however, account for more than a fraction of the diversity found in the eastern American Oenotheras. Numerous mutations must also be involved, and in this sense real evolution has occurred. Certain differences have also developed in the cytoplasm, and as a result the chloroplasts are affected so that pale green or colourless blotches may occur on the leaves in some hybrids. The chloroplasts in development may also have undergone changes of a mutational nature. The frequent occurrence in species hybrids, of lethal yellow seedlings which die in a week, is significant as showing that lethal as well as viable combinations can occur in a species cross. Everything goes to show the great range of unstable variability of recent origin. The advent of man and the cutting of forests may have much increased the variability by bringing together forms which were previously isolated, thus making hybridization more frequent. Colonies are sometimes found containing four or five distinct forms (see GATES 1936, p. 350).

Gene mutations in Oenothera

That mutations have occurred in the past can be deduced from the occurrence of gene mutations in cultures. Those from *O. Lamarckiana* and its derivatives, include *rubricalyx* (GATES 1911), old-gold[4] and *funifolia* (SHULL 1921), *supplena* (double flowers, SHULL 1927), *bullata* (SHULL 1928a), *pervirens* (SHULL 1932), *acutifolia* (BRITTINGHAM 1931), *stenophylla* and *angustifolia* (DE VRIES 1929). These two narrow-leaved mutations are phenotypically alike. They differ mainly in that in *stenophylla* the gene for small leaves lies in the *laeta* gametes. Its leaves are shorter than in *angustifolia*, more hairy and with more anthocyanin. Both are isogamous, non-splitting, with 14 chromosomes. DE VRIES cites many other mutant types in the same paper, a mutation rate of $8 - 10\%$ being reached. Mut. *proxima* has lost the zygolethal of the *velutina* complex, so it shows 96% of germination, producing two types, *proxima* and *retardata*. *Pustulata* occurred only once, and from 17 fruits only 14 seeds germinated, which belonged to the mutants *albida*, *cana* and *rubrinervis*. The original plant was thus nearly self-sterile, but this was not due to sterility alleles. Mut. *acutifolia* (BRITTINGHAM 1931) appeared in 1929 in SHULL'S cross-bred line of *Lamarckiana* which had been maintained for 23 years. The rosette leaves are narrower,

---

[4] Old-gold *(vetaurea)* and *sulfurea* petal colour are complementary (SHULL 1926), giving F_1 ordinary yellow, F_2 producing the double recessive segregate, *aurata*. The recessive gene *v (vetaurea)* was found to be independent of *br (brevistylis)*.

more sharply pointed and less crinkled. It is a simple monohybrid recessive. It grows more slowly and blooms several weeks later than the type and it retains the same catenation as *Lamarckiana*, the gene involved not being in SHULL's linkage group I.

These gene mutations are of interest as affecting leaf shape in a way similar to that of some recognized species. By the action of X-rays on the pollen of the homozygous mutant *O. blandina*, CATCHESIDE (1937) obtained three recessive gene mutations affecting leaf shape, i.e., b u c k l e d , h e l i c o i d and n a r r o w l e a f, as well as three g l o s s y triploid plants among 69 normals. The helicoid mutation bred true but the buckled could not be induced to flower.

Mut. *helix* from *O. (Lamarckiana X atrovirens) velutiflexa* was propagated by grafting on *O. biennis* and studied by CHROMETZKA (1955). The helicoid leaves are believed to contain an excess of growth substance.

In various Oenothera hybrids 37 new (named) genes are described by HARTE (1948) in a list of 59 genes affecting 34 characters, and 3 lethals. She concludes that nearly all characters in Oenothera are polymeric but this is an exaggeration.

As regards mutations in floral structure, SHULL obtained several in cultures of *O. Lamarckiana* which have some phylogenetic significance. It may be that his method of breeding, by intercrossing different individuals rather than selfing single plants, accounts for the occurrence of these mutants in his cultures and not in others, since a more heterozygous condition is maintained by his method. By crossing mut. *supplena* with mut. *brevistylis*, SHULL (1928b) obtained plants with flowers having no gynaecium, the ovary and hypanthium being replaced by a solid pedicel; but there were repetitions (up to four) of the successive whorls, – calyx, corolla, androecium. The functional stamens were usually confined to the outer one or two sets of whorls. Presence of the zygotic lethal affected the expression of these genes. Thus in *supplena-brevistylis* early flowers were normal, later ones having extra petals, style ± split at the top and shortened, the stigmas "massed together into a clavate structure" which is functional, although the style is covered with long hairs. In *decipiens supplena*, which has no zygote lethal, the carpels are so modified that the style and stigma are replaced by thin, plate-like "leaves", rarely functioning as ovaries. The extra petals in *supplena* vary greatly in number, most of the double flowers having four functional stamens and many petals. At the other extreme there are no extra petals, but the carpels are replaced by leaf-like structures. Thus both extreme forms of "doubling" are determined by a single gene. The significance of doubling as a parallel mutation in many other genera of wild and cultivated plants has been pointed out elsewhere (GATES 1920).

Mut. *pollicata*, from *O. Lamarckiana* (SHULL 1934) is of even greater morphological interest. In it the style, owing to its greater growth in thickness, is confluent with the wall of the hypanthium, leaving only eight small longitudinal air passages in cross-section of the flowerstalk. Pollen tubes function normally in passing down to the ovary, but the free style above the hypanthium is so slender and devoid of mechanical elements that it developes kinks and rests against the corolla like the clapper of an upturned bell. Both hypanthium and style are shorter than normal, and the stigma lobes are not spreading but more or less appressed and unequal.

This development of a solid flower stalk is reminiscent of features in some related genera of Onagraceae. It also occurred in a virescent strain of Oenothera from the Lancashire coast (GATES 1910a). The ring of 12 chromosomes remains unchanged in *pollicata*. Eight cases of this recessive gene mutation appeared in 1934, and its linkage relations were studied (SHULL 1937).

A trisomic mutation *(de Vriesii)* with flowers as small as *O. biennis* originated from *Lamarckiana* in 1915 (VAN OVEREEM 1922) and gave in 1916 a form *(bienniformis)* with 14 chromosomes and small flowers. Its complexes were *albigaudens-rubrivelans*. There is thus experimental evidence for small-flowered mutations.

Mutations in wild species

Since trisomic, tetraploid and triploid mutations occur with varying frequency in cultures, it is clear that they must also occur in the wild. But they have played no part in the evolution of Oenothera, since the wild forms all have 14 chromosomes[5]. The frequency of trisomics as mutations is undoubtedly connected with the existence of a ring of 12 or 14 chromosomes, in which alternate chromosomes are normally drawn to opposite nuclei in meiosis. A broadleaved *(lata)* mutant from *O. insignis* and trisomic *linearis* mutants from different strains of *O. ammophiloides* were recorded by GATES (1936), and GATES & NANDI (1935) described several trisomics from *O. paralamarckiana*. DE VRIES (1925) showed that many trisomic mutations, as well as plants with 16 or 17 chromosomes, are obtained in the offspring of triploids *(semigigas)*.

Numerous trisomics have since been analyzed and the extra chromosome in each determined as follows:

Trisomics	Species	Extra-chromosome	Author	Year
Lata	O. Lamarckiana, biennis, suaveolens	5·6	CATCHESIDE	1940
Dependens	O. Lamarckiana, biennis, rubricaulis	3·11	CATCHESIDE	1940

[5] CLELAND (1951) has found in two strains of *O. Hookeri* an extra pair of small chromosomes (very diminutive in one strain) as a relatively constant condition. Though very rare, this could be the beginning of a stable condition, but no effect of the extra pair (sometimes 3 or only 1) could be observed in the phenotype.

Scintillans	*O. Lamarckiana, biennis,*		10·13	RENNER	1943d, 1949
Cana	*O. Lamarckiana,*		3·4	RENNER	1949
Macilenta	*M-Lamarckiana*		2·3	RENNER	1949
Incana	*O. biennis*		4·9	RENNER	1949
Tripus	*O. Lamarckiana*		1·2	RENNER	1943d, 1949
Mira	*M-Lamarckiana*	2·3 + 1·11 +	1·2	RENNER	1949
Candicans	*O. Lamarckiana*	1·2 + 3·4 —	3·2	RENNER	1949
Glossa α	*albilaeta*	1·2 + 3·4 —	1·4	RENNER	1949
Glossa β	*albilaeta*	1·2 + 3·4 —	1·4	RENNER	1949
Paraglossa	*hookerirubata* 2 *albicans* chroms. present			RENNER	1949
	1 *hookeri* chrom. absent			RENNER	1949
Lonche	*albihookeri* *hookeri* 1·2 3·4 13·14			RENNER	1949
	albicans 1·4 2·14			RENNER	1949

Table I. Non–disjunction in various Oenotheras

Species	Catenation	Frequency (%)	Author
O. pratincola	14	3–7	KULKARNI 1929
O. franciscana sulfurea	12,2	16	CLELAND 1924
O. levigata var. *similis*	14	46	FORD 1936
O. ammophila	12,2	6	SHEFFIELD 1927
O. novae-scotiae	14	4	SHEFFIELD 1927
O. eriensis	–	7	SHEFFIELD 1927
O. Lamarckiana	12,2	10	CLELAND & OEHLKERS 1930
O. Cockerelli	–	11.7	CLELAND & OEHLKERS 1930
O. grandiflora	14	15.6	CLELAND & OEHLKERS 1930
O. grandiflora × *Lamk.* acuens.gaudens	14	15	CLELAND & OEHLKERS 1930
O. eriensis × *ammophila*	12	9	SHEFFIELD 1929
O. ammophila × *eriensis*	14	10–11	SHEFFIELD 1929
O. eriensis × *rubricalyx*	12	10	SHEFFIELD 1929
O. ammophila × *novae-scotiae*	14	10	SHEFFIELD 1929
O. ammophila × *rubricalyx*	6 + 4₁₁	8	SHEFFIELD 1929
O. rubricalyx × *novae-scotiae*	12	12	SHEFFIELD 1929
O. Berteriana × *O. mollissima*	14	40*	OELKRUG 1934
B. mK	· 14	30	OELKRUG 1934
l. mK			

* Also 3.3 % of 9 : 5 disjunction, one PMC 10 : 4 and 15 % with extra micronuclei.

It has been suggested by CLELAND (1944) that all the small-flowered forms might be grouped under the species *strigosa*, *biennis* and *parviflora*, but this would not be a taxonomic treatment. Four year's further growth of cultures on a large scale from new localities since the monograph (GATES 1936) was published, has led me (GATES 1957) to reduce to varietal rank certain of the forms originally described as species. This is not because they are indistinct or difficult to discriminate in cultures, but because they are obviously related to the forms to which I have attached them, as "varieties" differing from the species in a small number of characters. The majority of the new localities have yielded forms which fell clearly into the specific categories previously created. They confirmed the judgements already formed regarding most of the species. Other localities yielded forms which differed in certain characters only from species already defined.

Two collections presented characters which are new or rare in the genus. These are described as new species (GATES 1950, 1951a). In a few other cases forms showed relations with two or more previously described species. They have probably arisen at some time through crossing, but the evidence of hybridization is not always clear and they are more difficult to deal with from a taxonomic point of view. The question of hybridization is especially confusing because the two complexes which each type contains may be quite different, and most of these new species have not yet been analyzed into their complexes. Numerous crosses which had been made with this end in view could not be grown because of the war, so that a large amount of potential hybrid analysis has been lost. However, the complexes of *O. eriensis*, *O. novae-scotiae*, *O. angustissima*, *O. nutans* and *O. pycnocarpa* were determined in an earlier paper by GATES & CATCHESIDE (1932).

BARTLETT (1914a) described in detail twelve microspecies of Oenothera, mostly from further south, in Maryland, Virginia, Tennessee and Connecticut. Some of these have been reduced to varieties (GATES 1957) on the basis mainly of BARTLETT's indications of relationships. In a monograph on the early Oenothera work, GATES (1915) briefly characterized the species then recognized, and their distribution in North America, citing numerous specimens from European herbaria. But a new taxonomic treatment of these specimens, based on modern knowledge, is required.

Mut. *funifolia* from *O. Lamarckiana* had narrow leaves with rolled margins, and peculiar outgrowths from the lower surface. It appears to be parallel to the mut. *formosa* which BARTLETT found in *O. pratincola* cultures. In mut. *pervirens* (SHULL 1932) there was complete loss of red pigmentation, but no other difference from *Lamarckiana*. It produces the usual mutations, such as *lata*, *oblonga* and *nanella*. The old-gold *(vetaurea)* mutation is possibly an allele of *sulfurea*. *Vetaurea* and *supplena* formed a third linkage group with probably *bullata* which appeared in 1925 in the 19th generation of SHULL's *Lamarckiana*. It is not clear how there can be three linkage groups when 12 of the 14 chromosomes form a single ring. *O. suaveolens* (DE VRIES 1918) produced the parallel mutations *lata*, *sulfurea* and *lutescens*, as well as *fastigiata* (with narrow leaves and erect flowers), *jaculatrix* (near linear leaves) and *apetala* (low stature, narrow leaves, some flowers in the half-race having one or more petals).

Other species in culture have also produced many significant mutations of BARTLETT (1915b) in *O. pratincola* were *nummularia, tortuosa, rubricentra* and *nitida,* their main peculiarities being indicated by the names. In *nummularia,* in addition to the orbicular seedling leaves there were differences in the stem-leaves, in the pubescence of the buds, and the four sepals remain united when the flower opens, instead of separating in pairs. In a previous paper (BARTLETT 1915a) on mass mutation in *O. pratincola,* in which up to 100% of the progeny could be mutants, the names *albicans, setacea, formosa, revoluta* and *gigas* sufficiently indicate the character of the mutations. In a strain of *O. pratincola* with a ring of 12 chromosomes and a free pair (KULKARNI 1929), the free pair was found to carry the genes for flat leaf and revolute leaf, each giving a Mendelian 3: I ratio. In the strain with a ring of 14, non-disjunction (6–8 separation of the chromosomes in pollen meiosis) was 3–7%. This condition was originally discovered in plants (GATES 1908) and later named in animals (BRIDGES 1913). Other records of its frequency are found in Table I (p. 23), which indicates the potential frequency of trisomics. From examination of this Table it appears that non-disjunction is more frequent in F_1 hybrids than in the parent species. SHEFFIELD (1929) showed that the chromosome arrangements before they separate indicate the occurrence of double as well as single non-disjunctions.

All the Oenotheras with a ring of 14 chromosomes will produce trisomics from time to time, and all may also give rise to rare triploid and tetraploid *(gigas)* mutations. If the latter should happen to occur in a location to which it is ecologically adapted, it might produce a local tetraploid species. This is the way in which tetraploid species have arisen in other genera, but in Oenothera the catenation results in irregularities and consequent sterility when tetraploidy occurs, with great reduction in seed production. This could only be overcome in a considerable period of freedom from competition. The change in the shape of *gigas* pollen grains might even be regarded as a generic character.

Regarding the origin of the 4n Oenotheras, two possibilities were originally recognized, (a) the fusion of two diploid gametes, or (b) a doubling of chromosomes in the fertilized egg. GATES (1909a) supported the second view and showed that *O. gigas* is a cell giant. More recently, by an analysis of *O. atrovirens* X *O. biennis* hybrids, RENNER (1939) has found that the 31 *gigas* plants arising in 1600 offspring were mostly diploid-tetraploid chimaeras, indicating that chromosome doubling had occurred in the embryo some time after fertilization, Triploid *(hemigigas)* mutations were believed to arise generally through fusion of a haploid pollen grain with a diploid egg. The first triploid was described by GATES (1909b), although it was assumed at the time to be a chance hybrid rather than a mutation.

RENNER (1939) finds triploids occurring in various hybrid progenies, in which it can be shown that they generally arise from an unreduced embryo sac. The constitution of the triploid hybrids shows, however, that some arise through dispermy and more rarely from a diploid pollen grain.

In a *O. stenomeres* BARTL. – a species with cruciate flowers – the striking mut. *lasiopetala* was recorded (BARTLETT 1915c). It was evidently a trisomic, as 60% of the offspring were *typica*, 40% *lasiopetela*. It had broader leaves, much more pubescence, a more persistent rosette and hairy petals. A *gigas* mutation with 28 chromosomes also occurred. The *sulfurea* "mutation", which was known in *O. biennis* in Holland as early as the "Hortus Cliffortianus" (1737), appeared again in 1914 in the F_2 of *O. biennis* X *franciscana*, and bred true for cream-yellow flowers (CLELAND 1924). This is probably not a primary mutation; its appearance in this case may have been concerned with the cross, and the origin of the *sulfurea* gene must have been long ago.

Mutations with a different significance were described in *O. Reynoldsii* by LaRue & BARTLETT (1918). In a retrogressive series, *typica* ⸻→ mut. *semialta* ⸻→ mut. *debilis*, the plants became successively smaller and less vigorous, with shorter capsules. There was no heterosis in crosses between these forms. Measurements showed that the changes were concerned not with size of the cells but with their number and arrangement. The number of ovules was the same in the capsules of all three forms, but fewer ovules matured into seeds. A weakness in development was accompanied by increased sterility. But mut. *debilis* produced a mutation, *bilonga*, which was obviously progressive. These vigorous plants produced ovaries with more ovules, so that the stout capsules reached a length of 64 mm., twice the usual length and much longer than in any other species.

Evolution of the Onagraceae

Much evolution within the Onagraceae has been retrogressive in character. Taking Oenothera as a starting point, the genus Gaura is closely related and has independently developed chromosome catenation (BHADURI 1941), *G. parviflora* having 7 pairs, *G. Lindheimeri* 7 pairs in some plants, ⊚ + 4_{11} in others, *G. biennis* ⑭. In Gaura the 4 carpels remain as in Oenothera, but are reduced in size, the fruits being indehiscent with only 1 or 2 seeds developing in each. This is structurally a large reduction.

One of the diagnostic features of the genus Epilobium is the development of a pappus on the seeds, which is absent in Oenothera. This presumably progressive feature gave a great advantage in seed dispersal, and apart from human activity Oenothera is confined to

America while Epilobium is almost world wide. In Fuchsia the sepals are petaloid and not reflexed, the flower otherwise much as in Oenothera but the stigma entire. These three genera all agree in general floral structure, having a 4-parted flower with two whorls of stamens. Jussiaea has made a positive advance to a 5-parted flower, a condition which occasionally (see p. 78) occurs in Oenothera. In Clarkia the flower is 4-parted but one whorl of stamens is suppressed and the petals are specialized in becoming ± 3-parted. Similar petal-shapes occur in certain Oenothera hybrids (GATES 1923). Ludwigia also has but one stamen whorl. The Lopezia flower shows further marked reduction, with 4 sepals, 4 petals and 4 carpels but only 1 stamen and a staminodium.

In Circaea the reduction proceeds to an extreme which, if carried further, must end in extinction. In this genus the hypanthium is replaced by a short solid stalk, from the top of which the style arises. The small fruit is non-dehiscent and covered with adaptational hook-shaped hairs. In these genera the whole plant shows great reduction from a biennial or perennial to an evanescent annual. In the small flower of Circaea, as LaRUE & BARTLETT (1918) point out, the carpels are 2 instead of 4, a change which occurs as a somatic mutation in *Oe. pycnocarpa*. *Circaea lutetiana* is a delicate ruderal plant with 2 sepals, 2 petals, 2 stamens, 2 carpels and 2 seeds, while *C. alpina* reaches the end of the line with only one seed per flower. The length of this reduction series – reduction in size, structure and function – is impressive. It represents a line of extreme reduction, which plants essentially ruderal and ecologically opportunist can take.

The Gongylocarpeae, related to the Gaureae, developed some extreme and bizarre features. *Gongylocarpus rubricaulis* from Jalapa, Mexico, is a glabrous annual with deciduous petals, the linear sepals only 4 mm. long, ovary 2- or 3-celled with 1 ovule in each cell, style short and filiform. The hypanthium, 4–10 mm. long, is adnate to branch and leaf petiole, so the fruit becomes concrete with these structures. SMITH & ROSE (1913) described a new genus, Burragea, the type species of which was named *Gaura fruticulosa* by BENTHAM. It differs from Gongylocarpus in being a bushy perennial with linear leaves and large, showy flowers; but the ovary, as in the preceding genus, is sunk in the leaf axil, becoming embedded but tardily breaking away. This fruit is moreover diamond-shaped, 2-celled and 2-seeded. BENTHAM supposed it must be abnormal, although no disease was visible. In this perennial shrub, the flowers have 4 sepals which are highly coloured, oblong, becoming reflexed as in Oenothera. The very slender hypanthium is much longer (15 mm.) than the sepals and petals (7 mm.) and has an annular disc at its apex. Of the 8 filaments, 4 are longer than the rest, and the stigma is capitate, at length bipartite. In addition to *B. fruticulosa*, which has glandular hairs, another species, *B. frutescens*, is glabrous, with broader linear leaves, very short hypanthium (2–2.5 cm.) and larger petals (12 mm.). Both species grow in the desert conditions around Magdalena Bay in Lower California. In this group floral reduction has proceeded partly in the same direction as in Circaea, one genus becoming a small annual, the other a shrubby perennial with floral reduction; but the indehiscent fruit embedded in the woody stem is a specialization presumably in relation to the desert conditions.

We may now briefly examine the relationships of the genera of Onagraceae from the cytological point of view. Following the grouping of genera in RAIMANN'S monograph of the family in "Pflanzenfamilien" (1895), the chromosome numbers so far determined are as follows:

Table II.

Chromosome numbers in Onagraceae.

Jussiaeae		*Onagreae*	
Jussiaea	n = 8	*Clarkia*[6]	5, 6, 7, 8, 9, 12, 14,
Epilobieae			17, 18, 26 (LEWIS & LEWIS 1955).
Zauschneria	n = 15		Basic number probably 7.
Epilobium	n = 18	*Eucharidium*	n = 7
Chamaenerion	n = 18	*Godetia*	n = 7 (3n, 4n), 9
Gayophytum	n = 11	*Oenothera*	n = 7
Gaureae			
Gaura	n = 7	*Anogra*	n = 7
Fuchsieae		*Megapterium*	n = 7
Fuchsia	n = 11 (3n, 4n), 14.	*Taraxia*	n = 7
		Lopezieae	
		Lopezia	n = 11
		Circaeeae	
		Circaea	n = 11

It appears that 7 is the basic number for the family, being known in seven genera belonging to the Onagreae and Gaureae. The ancestors of the Fuchsieae, Lopezieae and Circaeeae, together with Gayophytum, evidently started off with 11 as haploid number, probably as the result of some revolutionary changes in nuclear structure. Certain species of Fuchsia have become tetraploid on this basis, some large-flowered horticultural hybrids have become octoploid, and one species has changed to n = 14 (WARTH 1925).

It is a curious fact of morphology that in Oenothera diploid plants

---

[6] VASEK (1955) has produced tetraploids (2n = 36) in *Clarkia unguiculata* by the use of colchicine. 4n × 2n gave triploids, and 3n × 2n produced aneuploids, some of which were double trisomics with 2 n = 20. LEWIS & ROBERTS (1956) find that *C. lingulata* (n = 9) differs from *C. biloba australis* (n = 8) only in the degree of lobing of the petals, its additional chromosome being homologous with parts of two chromosomes in *C. biloba*. Thus the origin of *C. lingulata* probably involved a paracentric inversion followed by the production of a constant tetrasomic.

have 3-lobed pollen while in tetraploids it is typically 4-lobed, whereas in Fuchsia haploid pollen is 2-lobed and diploid pollen 3-lobed[7]. An understanding of how a primary increase in cell size results in a characteristic change of cell shape remains for the future.

In Clarkia the evidence indicates (from the figures of HÅKANSSON 1931) that the change from 7 to 9 chromosomes took place by the fragmentation of two pairs, probably by misdivision at the centromere. In *Clarkia elegans* one plant had 9 free pairs of chromosomes while two others had a ring or chain of 4, again showing the beginning of catenation. In *C. pulchella*, with $2n = 24$, 6 to 10 of these chromosomes appear short, indicating again an increase in number by chromosome breakage. As mentioned previously (p. 13), the first four genera of the Onagreae have all developed catenation, probably independently. The cytological evidence is insufficient to indicate how Epilobium increased its chromosome number. In Jussiaea, the change to a 5-parted flower was accompanied by an increase from 7 to 8 pairs of chromosomes. Thus the cytological evidence helps considerably to clarify the relationships between the various genera. How old these genera are is unknown, but one may assume that the changes in basic chromosome numbers probably took place at least as early as the Pliocene.

In a series of careful taxonomic studies of the Onagraceae, MUNZ (see 1937), who has the characteristics of the lumper strongly developed, reduces to subgenera of Oenothera the following, many of which are regarded by some botanists as genera: Chylisma, Sphaerostigma*[8], Taraxia, Eulobus, Salpingia*, Megapterium*, Anogra*, Pachylophis*, Hartmannia*, Gauropsis, Gayophytum, Raimannia and Kneiffia. Many botanists will agree that such groups of species as Anogra and Raimannia should be treated as subgenera, but many will probably wish to regard Kneiffia and Pachylophis or Megapterium as genera. It is not to be expected that complete unanimity can be reached in such matters. In these three genera the fruits are

---

[7] The pollen of Onagraceae is easily recognized. Pollen grains most nearly resembling Jussiaea have been found in the Brandon (early Tertiary) lignite from Forestdale, Vermont. Very similar or identical grains have been found by THIERGART in the Oligocene of Germany, and onagraceous pollen has also been reported by RUDOLPH from Czechoslovakia. I am indebted to Dr. Alfred TRAVERSE of Harvard University, for these (partly published) observations and records. Since Onagraceae existed in the early Tertiary of Vermont, it follows that members of this family probably advanced and retreated at least four times during the Pleistocene glaciations in North America. Each such glacial episode probably had its effect, at least in extinguishing many species.

[8] Self-sterility is known (HAGEN, 1950) in the subgenera marked with an *. CROWE (1955) finds S factors in 4 species of Anogra, two of which have $④ + 5_{11}$. Another species is self-fertile. The S locus is shown to be homologous in all 4 species.

very different from the Oenothera capsules and from each other. In Kneiffia they remain small and become ribbed. The other two genera show their early origination by having very large flowers. In Pachylophis the fruits become large, woody and nodular, in Megapterium they develope extensive wings. Pachylophis species have become caespitose, forming flowers and fruits in the axils of the rosette leaves, and Megapterium approaches the same condition.

Fig. 2. Seedlings of 39.37.

Fig. 3. Seedlings of 51.37.

TAXONOMY

Oenothera Cultures

The original voluminous notes of the four year's work, 1936-1939 inclusive, involving hundreds of cultures, especially from Eastern Canada, were lost in the war (the notebooks for 1938 and 1939 have since been recovered), but many of the results were already analyzed and tabulated in records which have been preserved. The great majority of the cultures agreed with species or varieties already described in the previous monograph, but some differ in minor characters which will be pointed out. Thus it is now possible to determine to a large extent the geographic ranges of the various types (maps 1 and 2). The original localities for the new species and varieties are given in GATES (1936, Table XXVII, p. 348).

In order to show how early in ontogeny the differentiation between these species begins, 45 photos were taken of cultures in the seedling stage. Of these, 32 were taken on 15 April, 1937, 12 in 1938 and others in 1939. Only the culture numbers of these are known, and since the notebooks were lost, the species to which they belong cannot now be determined, except in two cases, Fig. 4 *(O. Hazelae)* and Fig. 5 *(O. ammophiloides)*. Figs 1 to 5 are included to show the range of variation between cultures and the uniformity within each culture. For Figs. 1 to 3 the seeds were all sown on the same day in February, 1937, and the seedlings photographed on the same date in April. Although of the same age, they show large differences in rate of growth and development, although grown under uniform green-

Fig. 1.   Seedlings of culture 40.37.

Fig. 4.   Seedlings of *O. Hazelae*.

Fig. 5.   Seedlings of *O. ammophiloides*.

house conditions. It will also be seen how greatly the leaf shape differs from one culture to another. A full study of seedling conditions might thus serve to differentiate species within the complex. This should convince taxonomists that a number of these forms are "good species", even to the lumpers.

Over 200 cultures of Oenothera were grown at the Courtauld Laboratory, Regents Park, London, in each of the last four years before the war. Some of these were first crosses and some were later generations of species already described, but many were from seeds collected in new localities. Since these mostly agree, or nearly agree, with the forms previously described, they fill in gaps in distribution and at the same time confirm the previous descriptions. Simultaneously, nearly all the previously described species and varieties were also grown, so that the living plants from new localities could be compared in every detail and every stage of development with the old forms growing in adjacent rows.

In order to systematize the work still further, a list of observations and measurements to be made on each culture was drawn up with the aid of Dr. T. A. Sprague and Mr. N. Y. Sandwith of the Royal Botanic Gardens, Kew. This underwent various changes, some

KEY TO ACCOMPANY MAPS

O. novae-scotiae	1	var. *niagarensis*	15b	*O. Jamesii*	34
var. *intermedia*	1a	*O. deflexa*	16	*O. organensis*	35
var. *distantifolia*	1b	var. *bracteata*	16a	*O. macrosceles*	36
var. *serratifolia*	1c	*O. pycnocarpa*	17	*O. Heribaudi*	37
O. comosa	2	var. *parviflora*	17a	*O. Hookeri*	38
O. Hazelae	3	var. *cleistogama*	17b	var. *angustifolia*	38a
var. *parviflora*	3a	*O. nutans*	18	var. *hirsutissima*	38b
var. *subterminalis*	3b	*O. atrovirens*	19	var. *irrigua*	38c
O. grandifolia	4	*O. cruciata*	20	var. *Hewetti*	38d
O. ammophiloides	5	var. *sabulonensis*	20a	var. *parviflora*	38e
var. *flecticaulis*	5a	var. *stenopetala*	20b	var. *Simsiana*	38f
var. *laurensis*	5b	*O. paralamarckiana*	21	var. *franciscana*	38g
var. *parva*	5c	*O. Nobska*	22	var. *venusta*	38h
var. *angustifolia*	5d	*O. rubescens*	23	var. *grisea*	38i
O. sackvillensis	6	*O. Oakesiana*	24	*O. longissima*	38j
var. *Royfraseri*	6a	var. *Tidestromii*	24a	var. *Clutei*	38k
O. leucophylla	7	*O. Shulliana*	25	*O. Hookeri*	38l
O. biformiflora	8	*O. canovirens*	26	var. *Wolfii*	38m
O. Victorini	9	var. *cymatilis*	26a	var. *Montereyen-*	
var. *parviflora*	9a	*O. chicaginensis*	26b	*sis*	38m
var. *intermedia*	9b	*O. furca*	27	*O. Reynoldsii*	39
var. *undulata*	9c	*O. disjuncta*	28	*O. pratincola*	40
O. apicaborta	10	*O. insignis*	29	var. *Numismatica*	40a
O. angustissima	11	*O. perangusta*	30	*O. syrticola*	41
var. *quebecensis*	11a	var. *rubricalyx*	30a	var. *litorea*	41a
O. argillicola	12	*O. strigosa*	31	var. *rhodoneura*	41b
O. grandiflora	13	var. *albinervis*	31a	*O. gauroides*	42
var. *Tracyi*	13a	var. *procera*	31b	var. *brevicapsula*	42a
O. levigata	14	var. *cheradophila*	31c	*O. ruderalis*	43
var. *scitula*	14a	var. *Cockerelli*	31d	*O. parviflora*	44
var. *similis*	14b	var. *subulifera*	31e	*O. rhombipetala*	45
var. *rubripunctata*	14c	var. *depressa*	31f	*O. heterophylla*	46
O. eriensis	15	*O. rubricapitata*	32	*O. Drummondii*	47
var. *repandoden-*		*O. MacBrideae*	33	*O. humifusa*	48
tata	15a	var. *ornata*	33a	*O. laciniata*	49

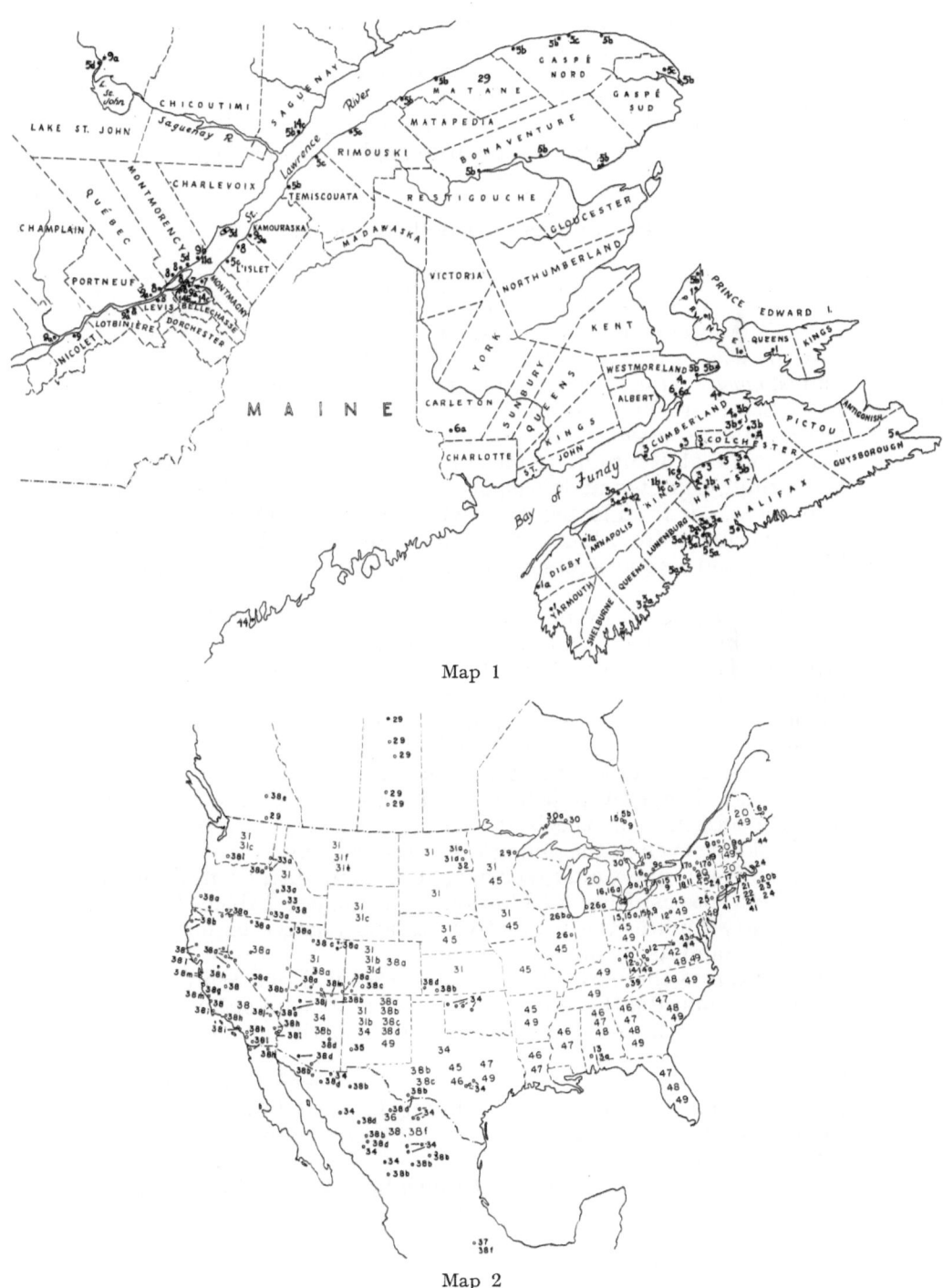

Map 1

Map 2

measurements and observations (e.g. apex of inflorescence, bracts, basal branches) being discarded either because they were not diagnostic or were too subject to environmental conditions or were not accurately measureable. The character of the branching was found to be of little value except in a few cases, as much depends on the conditions of growth – temperature, light and moisture.

Distribution of Oenothera species

Localities for each species, from garden cultures and from specimens in the Gray Herbarium, have been recorded elsewhere (GATES 1957). The species and varieties are listed by numbers in maps 1 and 2, where each is represented by a number. From these we may note that *O. novae-scotiae*, originally described from the Annapolis Valley, Nova Scotia, is found in a closely similar form in Prince Edward Island. Its var. *distantifolia*, previously from Newport (Hants Co.), is also found at Kentville (Kings Co), N.S. A specimen in the Gray Herbarium, collected at Kingsport, N.S., 1901, is identified as var. *serratifolia*, one from Meteghan as var. *intermedia*. These and many other specimens from Eastern Canada were collected by Prof. M.L. FERNALD and his colleagues in their study of the flora over a series of years.

O. Hazelae, originally described from Lockeport, Shelburne Co., N.S. has since been grown from many localities. It is one of the most distinctive of all the forms in culture. A specimen in the Gray Herbarium from Port Mouton, Shelburne Co., 1920, belongs to this species, but is naturally smaller in leaf dimensions than the cultivated form. The var. *parviflora*, originally from Port Mouton on the south coast and Middleton in the Annapolis Valley, has been extended to Port George (Annapolis Co.) on the Bay of Fundy coast and to three localities on the south coast. This very neat species is unmistakable even in the young rosette stage. Any taxonomist who observed it growing would recognize it as quite distinct from other Oenothera species. These 24 additional cultures (see Table III, p. 66) abundantly confirm the distinctiveness and relative uniformity of the species and its small-flowered variety.

O. Hazelae is the most widespread species in Nova Scotia, occupying nine counties. It is a handsome species of small stature, easily recognized, with liver-coloured spots on its smooth, subentire leaves; compact inflorescence; petals (including the varieties) 8-20 mm. long; and hooded sepals. The catenation is ⑭, whilst in *O. Hazelae* × var. *parviflora* it is ⑩ $+ 2_1$ (JACOB 1940), indicating that the species and its variety have only two chromosomes in common, i.e., with like ends.

O. ammophiloides extends from Lunenburg Co. to Guysborough,

N.S., and var. *laurensis* up the Gulf of St. Lawrence coast, around Bay Chaleur and the Gaspé peninsula and up the St. Lawrence River to Matane; also on the north shore of the St. Lawrence at Les Escoumains, up the Saguenay to Dolbeau on Lake St. John (as var. *angustifolia*) and at Lake Temiskaming on the western borders of Quebec (see map). Gray Herbarium specimens show that it is also in the western part of Prince Edward Island, a nearly related form occurring in Newfoundland. *O. flecticaulis* GATES has been reduced to a variety of *O. ammophiloides*, and its range extended from the mouth of the LaHave River to Mill Cove (Lunenburg Co.), Nova Scotia. This species in thus confirmed as a coastal and near-coastal form from the south shore of Nova Scotia up the Gulf coast and around the Gaspé peninsula, one variety being recognized on the south coast of Nova Scotia, another on the St. Lawrence Gulf, and the third (var. *parva*) on the south bank of the St. Lawrence from its mouth nearly up to Quebec. This is in part an ecotype living under more severe conditions.

From GATES (1957) it will be seen that many of the species described by BARTLETT and myself from different localities, as well as those of other authors, have been reduced to varieties. This has been done wherever the species to which they could be attached is fairly clear. This certainly helps in indicating relationships, the forms involved usually belonging to adjacent or the same areas.

The term variety as used here necessarily differs to some extent from the general conception, owing to the complexity of Oenothera genetics and the fact that in most of these forms the genetics has not been worked out. In short, these varieties are not derived from the species merely by one or more mutations, but there has probably been in many cases a redistribution of some of the linked genes in certain complexes of the species, so that by interpollination forms with several "linked" differences have arisen, the origin of the mutational differences being more remote. A species and its variety are also frequently found in the same area.

The following comments relate to the distribution of these forms. *O. novae-scotiae* is found in the Annapolis Valley and on the adjacent North Mountain. It extends west as var. *intermedia* (reduced from specific rank), and two other local varieties (all apparently supplanting the species) are found further east. The cultures from two localities in Prince Edward Island are inseparable from *O. novae-scotiae* and are nearly indistinguishable from it, but the rosette leaves are less crinkled. Compare Figs. 5 and 6 in GATES (1936).

When the larger-flowered valley and smaller-flowered mountain strains of *novae-scotiae* were crossed with *O. Lamarckiana* pollen, the difference in catenation of the F_1 (⑩ or ⑧, PATHAK 1940, see Table IV, p. 69) indicates a difference of one chromosomal interchange between these two strains.

$O.$ *comosa* has characters belonging to both $O.$ *novae-scotiae* and
$O.$ *Hazelae* (GATES, 1936, p. 265). Its origin probably involved
crossing between certain strains of these two species, so in this sense
it is a species of hybrid origin, breeding true because of its ring of 14
chromosomes. Both the presumed parent species occur in the locality
where it is found, but its range of distribution is unknown. Such a
hybrid species, on the conceptions adopted here, does not necessarily
consist entirely of the descendants of a single cross. Rather, various
crosses may have occurred in different localities between different
strains of the two parent species. One must similarly assume that in
the post-glacial history of many other species a complicated net-
work of occasional or rare crosses has taken place, leading to the
numerous cross-relationships which can be observed.

$O.$ *grandifolia* is easily recognized in the field by its large rosettes
of very broad leaves, short stem with no red papillae, and short,
appressed, terminal sepal tips. There are indications that one of its
complexes may be similar to one of $O.$ *novae-scotiae*, although phenotypically the two species are extremely unlike. It is at present known
only from Colchester and Cumberland Counties, N.S., and West-
moreland Co., New Brunswick, but may be more widely distributed
in eastern Nova Scotia. Unlike other forms, it appears to be mainly
inland. A culture of $O.$ *grandifolia* in 1936 produced a *gigas* mutation. From its seeds 30 plants were grown in 1938, the rosettes
showing the same wide range of leaf width as in $O.$ *Lamarckiana*
gigas. They only came into flower the following year; petals 19–26 x
22–28 mm.

The distribution of $O.$ *ammophiloides* with its three varieties
(two of them reduced from species rank) has already been referred
to. Its oblique stems with strongly bent tips recalls the *percurvans*
complex of $O.$ *ammophila* FOCKE, naturalized in Europe; but the two
species are not otherwise closely related. When two different plants
were crossed with $O.$ *Lamarckiana* pollen, the F_1 had ⑧ + ⑥ in one
case and ⑧ + ④ + 1_{11} in the other (PATHAK 1940), thus indicating one
interchange in plants which were phenotypically alike.

$O.$ *sackvillensis* and its varieties are at present known only from
one locality in New Brunswick, but the Gray Herbarium has a
specimen representing var. *Royfraseri*, collected at McAdam
Junction, N.B., in 1916. $O.$ *leucophylla* is known from two adjacent
counties on the south bank of the St. Lawrence near the tidal limit.
$O.$ *biformiflora* is found in this region on both banks at many localities.

$O.$ *Victorini* with its three varieties occupies a wide area from
Kamouraska Co. far below Quebec City, along both banks of the
River to Montreal; also from localities in northern Quebec. In
recognizable form it extends west to Toronto and Essex Co.,
Ontario, and southwards into northern New York State.

O. angustissima was originally described from Ithaca, New York, its var. *quebecensis* from Cap Tourmente (Montmorency Co.), Quebec. Its complex-constitution was afterwards determined (GATES & CATCHESIDE 1932) as *divergens (rubrans)* ♀. *divergens* ♂. *Rubrans* contains one or more genes for larger flowers and for the diffuse red colouration. The cross *O. angustissima* × *eriensis (rubrans undulans)* scarcely differs from *angustissima* except in the bent stem tip. The complexes of *angustissima* and of *novae-scotiae* both appear to differ in genes for flower-size, but since flower-size is partly cytoplasmic in its determination (p. 93) this matter requires further investigation. As regards relationships, the *divergens* complex of *O. angustissima* produces narrow leaves, while both complexes of *O. eriensis* have narrow leaves, but otherwise these two species are very different. Again the bent stem determined by the *undulans* complex of *O. eriensis* is like the crozier stem of the *percurvans* complex of *O. ammophila*, but the home of the latter species is on the Atlantic coast and there is no other indication of relationship to *O. eriensis*.

O. angustissima var. *quebecensis*, from the north bank of the St. Lawrence (GATES 1936, p. 324), is clearly related to the species in being glabrate, with deep red stems and midribs, narrow leaves, nutating stem tips and markedly subterminal sepal tips. This last feature, with the rather thin and wiry branches, suggests the subgenus Raimannia. The catenation of the F_1 hybrid *O. angustissima* × var. *quebecensis* is ⑧ + ④ + 1_{11} and of the reciprocal ⑧ + 3_{11} (JACOB 1940), thus confirming their relationship. The light-sensitive papillae on ovary and hypanthium recall *O. ammophiloides*, and the liver spots on the leaves are conspicuous also in *O. Hazelae*, and in *O. nutans* from Ithaca, New York.

O. argillicola & *O. levigata* were originally described from White Sulphur Springs, W. Virginia. The former species is known also from Virginia, Maryland, and New York. *O. scitula* BARTL. from White Sulphur Springs, has been reduced to a variety of *O. levigata*, and two other vars. of this species are recognized from St. Vallier, Quebec.

O. eriensis GATES is now known from Essex Co. (GATES 1927a), Wasaga (cult. 229. 39)[9] and Collingwood (Fig. 6, cult. 226. 39) in Ontario, and a closely related form from Lake Temiskaming, Quebec Fig. 7, cult. 42. 39.

This species is even more prone to omit the rosette than is *O. grandiflora*, to which it appears to be wholly unrelated. The Wasaga Beach culture (229. 39) was very tall, stem tip slightly bent, papillae rare,

---

[9] This means culture 229 in 1939. If one of these plants had been selfed and its seeds grown in the following year, it might have been numbered e.g. 106.40. This will explain the culture numbers used in this paper.

Fig. 6.   *O. eriensis*, Collingwood (Grey Co.) Ontario. Culture 226.39.

white or pink; leaves very narrow, smooth repand-denticulate, midribs white; inflorescence dense, bud cone green, no papillae on ovary;
sepal tips stout, terminal, not in contact. Cult. 220. 39 from Longueil
(Chambly Co.) Que., is closely similar in measurements, except
height, but differed in (1) erect stem, (2) minute pink papillae on
ovary, (3) sepal tips subterminal, ± erect prongs tipped with red.
One may safely conclude that any two forms which so nearly agree

Fig. 7. *O. eriensis*, Lake Temiskaming, Quebec. Culture 42.39.

in all these measurements will differ only in minor phenotypic characters. The cultures 41.38 and 42.39, both from Lake Temiskaming, agree closely in measurements except in height. They are distinguished from *O. eriensis* only by their entire leaves, which are green (not grey-green). The flowers are larger in certain measurements and the sepal tips shorter.

The complexes of *O. eriensis* are *glaucens* • *undulans*, resembling respectively *rigens* and *percurvans* of *O. ammophila* (GATES & CATCHESIDE 1932), but in *eriensis* both types of eggs are functional. The indications from the catenation in F_1 hybrids of *O. eriensis* with other species are that its segmental interchanges have followed a different course from those in several other species. Thus *O. erien-*

42

sis × *ammophila* gave ⑫, the reciprocal, ⑭ and *O. eriensis* × *rubricalyx* ⑫ (SHEFFIELD 1929). The cross *O. rubricalyx* × *eriensis* was made four times in 1927 and 1928, producing a total of 146 seedlings nearly all of which died in the cotyledon stage. One which survived proved to be a haploid (GATES & GOODWIN 1930). The *eriensis* pollen had stimulated a *rubricalyx* egg cell to develope parthenogenetically. Later this cross was repeated 16 times, and from nearly 1000 seedlings obtained 38 plants survived the seedling stage, all of which were apparently hybrids. The cross *O. rubricalyx* X *eriensis* is thus of special interest. The few which survive the cotyledon stage form a continuous series in size, strength and development. Only one plant in 85 was fully green and of normal size. The catenation (HEDAYETULLAH 1932) was generally ⑩ + ④ or ⑧ + ④ + ②.

The vars. *repandodentata* and *niagarensis* of *O. eriensis* were reduced from specific rank owing to the obvious near relationship to the species. This extends the range of the species along the length of Lake Erie and northwards to Lake Temiskaming, Que. The relationship between species and varieties is indicated by the cross *repandodentata* X *eriensis* having ⑩ + ④ (JACOB 1940).

O. deflexa occurs from Windsor and Ojibway, Ont. to Grenville (Argenteuil Co.), Montreal, and Quyon on the Ottawa River (Temiskaming Co.), the var. *bracteata* being known only from the original locality at Windsor. The strain from Ojibway, Ont. (cult. 144.38), agrees closely in measurements with the species, but differs in (1) taller stem, (2) no red papillae.

O. pycnocarpa is known from cultures from Ithaca, N.Y., Vineland, Ont., and Long Island, N.Y., and herbarium sheets from New York, Connecticut, Pennsylvania and West Virginia (GATES 1957). Its complexes were *dependens (dentans)* ♀ • *dentans (dependens)* ♂. *O. pycnocarpa* and *O. Victorini* appear to run into each other in northern New York State.

O. pycnocarpa vars. *parviflora* and *cleistogama* are from seeds collected at three other localities in New York State by Dr. G. L. STEBBINS Jr. *O. nutans* was also described from Ithaca, N. Y., but is recognized in herbarium sheets from Maryland and Virginia. Its complexes are *serratans* ♀ • *nutens* ♂. Phenotypically *pycnocarpa* and *nutans* are nearly related and sharp distinctions are difficult to find. They were also described from the same locality and the F₁ hybrid combinations *serratans · dentans* and *dependens · nutens* are closely similar (GATES & CATCHESIDE 1932), yet they breed true in F₂ and have a ring of 14 chromosomes like the parent species. This might be because their segmental interchanges have all been different, so that the contrasting features of the *nutans* complexes are differently arranged in the *pycnocarpa* complexes. Thus the differential genes in these two microspecies appear to be few, but despite their near relation they may have no whole chromosomes in common. The ⑭ in the hybrids cannot be used as a sign of distant relationship. This condition could imply that intercrossing has been rare in the past. On the other hand, it may mean (and this is more probable) that all the chromosome-ends of both species are the same, in which case these two "species" would simply represent segregates of a rather freely interbreeding population.

O. paralamarckiana and O. Nobska are from different areas near Woods Hole, Mass. As O. Lamarckiana forms were grown at the Marine Biological Laboratory in the years 1905–08, it is probable that the former species derived its Lamarckiana-like characters from pollen from that source (GATES & NANDI 1936). It produced various trisomic mutations. O. Nobska (STURTEVANT 1931) is now reduced to a variety of O. Oakesiana. It has the complexes pubens ♀ aenescens (♀) ♂. Its phenotype is most like O. Oakesiana, having narrow, greyish-green leaves with red midribs, but no red papillae.

Cultures of O. Oaskesiana and O. Nobska were grown side by side in 1938 from seeds collected respectively at Falmouth, Massachusetts and Nobska Point (a few miles away) by STURTEVANT & CATCHESIDE. In 1938, var. Nobska cultures differed only in (1) stem and midribs pink, (2) very small red papillae on the stem, which are rare in Oakesiana, (3) smaller petals (11-12 x 9 mm against 14 x 16 mm), (4) ovary and sepals much more pubescent with long hairs, (5) sepal tips 2 mm (against 3 mm), (6) shorter and stouter capsules. The stem is erect but ± wobbly in both. Strains of Oakesiana from Long Island differed in minor features, such as slightly broader leaves, straight stem, sepal tips less subterminal, red at base of hypanthium, leaves slightly darker green, petals slightly smaller (10 mm.).

Var. Ostreae has cruciate flowers and is reduced to a variety of O. atrovirens (GATES 1957), which has dark green, shiny leaves. Var. Ostreae usually forms no rosette. The leaves are narrow with red midrib, stem green with curved tip, often fasciated. The sepal tips are slightly subterminal, scarcely spreading, buds reddish throughout cone and hypanthium, as in O. rubescens from Nantucket Island. The fascians complex is in the egg cells.

O. Oakesiana occurs on the New England coast from Falmouth, Mass. to New Haven, Connecticut and Long Island, N.Y. The strain from Cold Spring Harbor showed the complexes acclerans ♀ denudans ♂ (STURTEVANT 1931). It differs from O. muricata conspicuously in having (1) no red papillae, (2) the buds and fruits glabrate. Its var. Tidestromii (reduced from specific rank) occurs further south in Maryland. O. Shulliana from Morristown, N.J., is tall and much branched, its complexes jugens ♀ (♂) maculans (♀) ♂. The leaves are moderately narrow with red midrib and "red flecks" (liver spots); stem "lightly punctate," erect; sepal tips terminal; buds yellowish because of the thin sepals.

O. canovirens in Illinois has blue-green foliage and petals 10-14 mm long. An eastern drift of some species is shown by its presence on Nantucket Island, at New York and at Lynbrook, Long Island (BICKNELL, 1914) as a natural element of the flora. It differs from *O. strigosa*, an apparent recent introduction on Nantucket, in having a longer hypanthium and longer sepal tips. This floral drift is confirmed by the presence of an Opuntia on Nantucket and formerly on Manhattan Island. *O. canovirens* var. *cymatilis* (reduced from a species) from Sawyer (Berrien Co.) Michigan, having similar foliage but larger petals (16 mm) is clearly related. *O. chicaginensis* (= *O. biennis* Chicago de Vries) has the complexes *excellens* ♀ *punctulans* ♂ (RENNER & CLELAND 1934). The catenation is constantly $\textcircled{12} + 1_{11}$. The petals are 21 x 25 mm but it is quite different from *O. biennis* or *O. canovirens*. The *excellens* complex has genes for (1) red midribs, (2) green papillae, whilst the *punctulans* complex includes genes for (1) smaller leaves, (2) white midribs, (3) small red papillae.

O. rhombipetala from Minn. to N.Y. and Texas has rhombic petals 18 mm long and pinnatifid rosette leaves[10]. It has seven free pairs of chromosomes and belongs to an independent line of descent and dispersion. *O. heterophylla* from Georgia to Texas (petals ca. 13-15 mm) is related. It also has pinnatifid rosette leaves. These two species belong to the subgenus Raimannia. *O. grandiflora* also has pinnatifid rosette leaves, but they only appear under conditions of slow development. It has been shown (HECHT 1944) that some strains of *O. rhombipetala* from Bridgeport, Indiana, are self-sterile and some self-fertile. By colchicine treatment of the self-sterile race a tetraploid strain was produced which was still self-sterile, probably through the original presence of a single pair of S alleles. Thus self-sterility has developed independently in this species and in *O. organensis* (see below) as well as in Megapterium.

O. Drummondii, which also belongs in Raimannia, has linear pods, and ranges from Florida to S. Carolina, Louisiana and Texas. In this southern area the petals range from 18-40 mm, frequently drying red, hypanthium 25-80 mm, ovary 8-14 mm. The leaves, upper stem and ovaries are densely hirsute, stem leaves very narrow (midleaf 28 x 10 mm). In the related *O. humifusa* NUTT. (New Jersey to Maryland, Florida and Alabama) the rosette leaves, and in *O. laciniata* HILL (Maine to Florida and Ohio to New Mexico) the stem leaves are pinnatifid.

O. strigosa is represented by Figs. 8 (rosette) and 9 (in flower) of typical plants of this widespread species in the Western States, as grown in Regents Park, London, from seeds collected at Barrie, N.

---

[10] A specimen from Oklahoma, 1939, in the Gray Herbarium has petals 20 mm long, and red spots on the sepals, as in some strains of *Megapterium missouriensis*. The petal length ranges from 12–25 mm. HECHT (1950) has made a full study of the subgenus Raimannia, especially in South America.

Dakota, in 1932. The soft-silky pubescence of the plant on the plains
has been lost in culture in the moister climate of England, but the
phenotype appears to be otherwise unchanged. Its varieties will
be considered below. Its complexes *(deprimans* ♀ *stringens* ♂
RENNER) were investigated by OEHLKERS (1923) in crosses with
O. Cockerelli and *O. suaveolens. Deprimans* is very like the *curtans*
complex of *O. Cockerelli. O. strigosa* covers a wide area from Minnesota to Washington State, Utah, New Mexico and Kansas.

O. strigosa var. *cheradophila* (reduced from a species) has petals
8 mm. or less in length and occurs in Washington State and Wyo-
ming. Var. *albinervis* (also reduced) has conspicuous crinkling, at
least as grown in England. It is known from three localities in North
Dakota. Other forms here grouped as vars. of *O. strigosa* are var.
subulifera from Montana with short sepal tips and petals 8 mm or

Fig. 8. *O. strigosa* rosette. Culture 76.33.

46

less, var. *procera* from New Mexico and Colorado, and var. *depressa*, a prostrate form from Montana, with broader leaves and much denser pubescence. *O. Cockerelli* BARTL. from Colorado (reduced to

Fig. 9. *O. strigosa* in flower. From Barrie, N. Dakota. Cult. 76.33.

a var. of *O. strigosa*) was extensively studied by OEHLKERS (1921) and has the complex formula *curtans* ♀ *elongans* ♂. It is strigose hairy, with stem 140-150 cm bearing red papillae, green sepals 36 mm, short style, petals 15-17 mm. *O. rubricapitata*, known only from one locality in North Dakota, is a very distinct and ornamental species with red shoulders to the buds and crinkled leaves, petals 13 mm.

The catenation of the F_1 hybrid *O. strigosa* var. *albinervis* X *O. rubricapitata* and its reciprocal is of interest. The strain of *albinervis* from Fargo, N.D. gave in this cross ⑫, while that from Barrie, N.D. gave ⑧, thus showing three pairs of chromosomes with the same ends as *rubricapitata*. In the reciprocal cross there were two F_1 types, the one with red papillae having ⑭ while the one without had ⑩ + ④ (JACOB 1940). A specimen in the Gray Herbarium from Leeds, N.D., 1898, agrees with the species *rubricapitata* except that there is no evidence of the peculiar bud colouration.

From a detailed study of many Western strains of *O. strigosa*, CLELAND (1954) traces lines of emigration and evolution within the species by following the various interchanges of chromosome ends which have occurred. The resulting complexes are nearly all at least 3 interchanges removed from a form, assumed ancestral, having ends 1·2 3·4 5·6 7·10 9·8 11·12 13·14.

O. insignis, considered more fully below, has a wide distribution, with little variability over the Canadian prairies and further east in Ontario and Quebec. Cultures have been grown from all the localities listed in GATES (1957).

Coming now to the large-flowered forms, *O. grandiflora* and its var. *Tracyi* (reduced from a species) both have a rather evanescent rosette unless grown in conditions which hold back development. When grown in such conditions the later rosette leaves are strongly pinnatifid at the base (GATES 1915, p. 13), but at higher temperatures the rosette is omitted altogether. *O. grandiflora* was rediscovered (MacDOUGAL, VAIL & SHULL 1907) at BARTRAM'S original locality in Alabama (VAIL 1905). Var. *Tracyi*, an annual with evanescent rosette, has the same early growth habit, and liver spots on the leaves. It differs mainly in having small flowers. A specimen in the Gray Herbarium from Berkeley Co., South Carolina, 1939, belongs to *Tracyi*. GERHARD (1929) found *O. grandiflora* to be heterogamous but isogamous, its formula *truncans* ♀♂ • *acuens* ♀♂, these terms referring to the flat or pointed terminal inflorescence. The complex *acuens* differs from *flavens* (of *O. suaveolens*) mainly in that *flavens* bears a zygolethal, while *acuens* does not. The combination *acuens· acuens*, known as *ochracea*, is therefore viable and occurs in about one-third of the offspring when *O. grandiflora* is selfed. The segregate *ochracea* differs from *O. grandiflora* only in quantitative characters and blooms ten days later. Being homozygous, it has 7 free pairs of chromosomes, while *grandiflora* has ⑫ + 1_{11} (GERHARD 1929), like *O. Lamarckiana*. However, CLELAND (1931) finds always ⑭ in the strains of *O. grandiflora* examined by him[11]. In the cross *O. grandi-*

---

[11] CLELAND (1950, p. 73) shows that in his original plant of *O. grandiflora* in 1936 with ⑭ a segmental interchange took place, changing 1.4 13.4 to 1.13 4.14 and so producing a constant strain with ⑫.

flora X *O. Hookeri* the F_2 showed seven different types in a small culture, indicating much segregation, i.e., many gene differences, but the catenation in F_1 was unfortunately not determined.

It seems evident that *O. suaveolens*, which grows widely in France (and also in Berlin, RENNER 1941) is a hybrid species; one parent could be *O. argillicola* (see below) and the other perhaps *O. biennis*. The Alabama species, *O. grandiflora*, is not very closely related to *O. argillicola* in W. Virginia or to *O. Hookeri* in California. These large-flowered species with little or no catenation represent an earlier (probably pre-Pleistocene) stage in the evolution of the genus.

O. MacBrideae from Idaho has petals 44 mm long, but its catenation is unknown. *O. ornata*, reduced to a var. of *O. MacBrideae*, is also from Idaho. It has petals 25 mm, presumably owing to the introduction of a dominant factor or factors for smaller flowers. *O. Jamesii* T. and G. from Texas and New Mexico, has petals 40 mm, hypanthium 70-110 mm. The enormously long hypanthium of this and other species seems impossible to account for on the principle of natural selection. MUNZ merges the species with *O. Hookeri* var. *hirsutissima* despite the big differences.

O. organensis MUNZ, from the Organ Mountains in New Mexico, is narrowly endemic. It was described as *O. macrosiphon* by WOOTON & STANDLEY, but this name was preoccupied. Its petals reach 50-60 mm. This species has been much studied as regards its catenation and especially its self-sterility factors, a new development in the genus which could hardly arise in the self-pollinated species. Self-sterility is now known in many Western species belonging to different subgenera. Not only the two Raimannias already referred to, but *O. (Megapterium) missouriensis* (GATES 1939, LINDER 1950) (which also has seven free pairs of chromosomes, HEDYATULLAH 1933), *O. organensis* (EMERSON 1938), as well as probably *O. (Anogra)) californica*, *O. (Pachylophis) caespitosa* var. *marginata* and *O. (Lavauxia) acaulis* are self-sterile. HECHT (1944) finds that the perennial species *O. latifolia* (RYDB.) MUNZ, *O. pallida* LINDL., *O. runcinata* (ENGELM.) MUNZ and *O. trichocalyx* NUTT. are also probably self-sterile. Thus while the subgenus Oenothera (formerly Onagra) has specialized in self-pollination and chromosome catenation, various other subgenera have retained open pollination and evolved self-sterility factors. *In O. missouriensis* the chromosome pairs have remained free, while in *O. organensis* they have developed catenation through interchange of ends.

O. organensis

O. organensis is confined to certain neighboring canyons in the Organ Mountains, New Mexico, where it grows on gravels derived from igneous rock (rhyolite) at an altitude of over 6000 feet. The entire population of the species, growing in three canyons under these specialized conditions, probably numbers not over 500 plants (EMERSON 1939). They are strictly perennial, having no central stem, but forming lateral branches from the axils of the rosette leaves. The roots are fibrous, with no tap root. The first plants were collected here in 1881, and WOOTON & STANDLEY described the species in 1913. Each canyon contains an inbreeding population, but 13 self-sterility (S) alleles were common to two canyons and 12 to two others, 45 S alleles being found altogether. EMERSON made tests of the pollen grains and found no S factor mutations in 45,000 grains. In a mathematical analysis of this situation, SEWALL WRIGHT (1939) concluded that a mutation rate of 10^{-5} would maintain less than 15 alleles in a general interbreeding population of 500-1000 individuals. If separated into 50 local groups of only 10 plants each, with only 2% of foreign pollen coming in, it would require a mutation rate of 9×10^{-5} to maintain 50 alleles. There must therefore be a very large amount of local pollination between neighbouring plants, although self-pollination produces no seeds.

EMERSON (1940) found that compatible differs from incompatible pollen only in the extent of pollen tube growth. There is no bud or end-season fertility, and inhibition of fertilization by incompatible pollen is complete. The inhibition of pollen tube growth is not governed by diffusible substances but is a reaction between the pollen and the stylar tissue. EMERSON (1941) showed that, as is usual with self-sterility, each plant is heterozygous for two S alleles. A pollen grain carrying any one of the 45 alleles fails to grow normally in the style of plants carrying the same allele. A pollen lethal factor was also found. Grains having this gene put out several pollen tubes but failed to germinate normally on any stigma. The sterility allele S_{20} was found to be associated (linked) with this pollen lethal, the cross-over value being $0.3 - 0.5\%$. LEWIS (1947) obtained tetraploid O. organensis by using colchicine. In the diploid pollen of such plants some allele pairs compete, resulting in compatibility of the pollen grains in styles carrying both alleles. The S alleles which do not show competition may show dominance of one allele over the other in the pollen grain.

O. organensis was found by EMERSON to be cross-fertile with five species of Raimannia, but hybrids with other Oenotheras "are obtained infrequently, and are almost completely sterile". On this basis it is more nearly related to Raimannia, but EMERSON suggests it should be placed in a new section between Raimannia and Oenothera. EMERSON (1938) found individuals with a ring of six chromosomes and others with a ring of four. The nucleolus was always associated with the ring, while the S locus was not in the nucleolar chromosome, and the male-sterility gene was independent both of S and of the ring.

STEINER & SCHULTZ (1958) have recently found incompatibility or S alleles in small-flowered species such as strains of *O. biennis*, *O. muricata* and *O. Victorini*. Selfsterility was discovered in *O. organensis* (EMERSON 1938) and later in the large-flowered *O. missouriensis* (GATES 1939). The presence of S factors is therefore probably an early condition which has persisted in the small-flowered species from their large-flowered ancestors. In *O. biennis* the α-complex carries an incompatibility allele (S_1) while the β-complex carries an allele which has no incompatibility effect. The S_1 in the α-complex thus serves as a pollen lethal, because pollen tubes carrying it are inhibited in the style by the presence of S_1. It appears therefore that the S genes in the small-flowered species, which may at times lessen their seed-production, have been inherited from an earlier stage in the evolution of the genus. In the large-flowered open-pollinated forms they functioned to maintain heterozygosity, but in the later small-flowered species they survive as a relic condition whose presence may be a disadvantage and cannot possibly be an advantage to the species.

O. macrosceles[12] from Northern Mexico has petals only 20 mm long, but hypanthium 90 mm and ovary 15 mm. Thus in this series (unlike the eastern forms) the petal size can be reduced while the hypanthium extends to its maximum length. In all these southwestern forms the great length of the hypanthium is a feature. As this means a much longer path for the pollen tube to traverse, some action must have been at work contrary to that which reduces the flower in all its other parts. Many of these forms may be self-sterile, and of course the longer the hypanthium the greater the chance for competition between pollen tubes.

The hypanthium is a characteristic floral organ in Oenothera and related genera. It is absent in certain genera of the Onagraceae and, as already mentioned, it reaches an enormous length in some of the large-flowered (more primitive) species. Under insect attack of the buds it may be suppressed (GATES 1910a), and also in a virescent race of *O. multiflora* from the English coast near Liverpool (GATES 1914a). SHULL (in MACDOUGAL et al. 1907) showed statistically that it is more variable than other organs of the flower. The subgenera Salpingia and Calylophis, of the genus Oenothera (treated as genera by some botanists) both have yellow flowers and a discoid stigma, but they differ mainly in that Salpingia has a long hypanthium, reaching 50 mm or more in length, while in Calylophis it remains short (MUNZ, 1929).

O. Hookeri

O. Hookeri extends from Mexico to California and S. Colorado,

---

[12] This species is placed by MUNZ (1935) in the subgenus Raimannia along with *O. rhombipetala* NUTT., *O. heterophylla* SPACH and *O. Drummondii* HOOK.

having petals *ca.* 40 mm long. CLELAND (1935) studied nine cultures of *O. Hookeri* from different habitats in California. They generally had ④ + 5_{11}, or 7_{11} chromosomes, but one plant had ⑧ and a few showed ⑥ or ④ + ④ + 3_{11}. Many plants were entirely homozygous, there were no lethals or gene complexes, and hybrids between strains usually gave plants with 7_{11}. These strains were all closely related and showed the beginnings of catenation.

O. irrigua and *O. Hewetti*, both from New Mexico, are (GATES 1957) treated as varieties, or derivatives of *O. Hookeri*, *Hewetti* having medium sized flowers (petals 30 mm). A culture of ten plants from seeds kindly sent from the original locality by the late Dr. T.D.A. COCKERELL, was grown by the writer at the university of California in 1915. They showed great uniformity, except that 6 had pink midribs and diffuse red stems, 4 white midribs and green stems. The petal size ranged from 25-33 mm with the mean at *ca.* 30 mm, a rather wide phenotypic range. All were strongly pubescent, with no red papillae on any part. The leaves were long and narrow, resembling the mutant *rubinervis* (from *O. Lamarckiana*) but scarcely crinkled. The styles were "long" and the buds reddish like *rubrinervis*. These observations were recorded in 1915. Measurements were made by COCKERELL (1919). He also records pistil length 78 mm (13 mm beyond the stamens, as in *rubrinervis*), plants large and spreading *ca.* 4 feet high. It would be interesting to know if this form has seven free pairs of chromosomes.

O. Hookeri var. *parviflora* is described on p. 12, based on specimens from British Columbia. Var. *angustifolia* (see GATES 1915, p. 30-31) differs from the type mainly in its narrower leaves (8-12 mm wide) and bright red stem. In addition to the type from Utah, 21 exsiccata were cited, some of which may belong to *O. MacBrideae* or its var. *ornata*. Many specimens in the Gray Herbarium from Utah, Idaho, Colorado and North Nevada also belong to this easily recognized form and have been referred to it by MUNZ. The petals range from 25 to 35 mm.

Var. *hirsutissima* (GRAY) MUNZ (l.c., p. 32) differs from *O. Hookeri* in (1) very short sepal tips (2 mm instead of 4 mm), (2) very long and loose pubescence on stem, hypanthium, ovary and sepals. It is widespread in the West, in Utah, Arizona, New Mexico, Texas and Chihuahua (Mexico). The petals run from 25 to 45 mm and they frequently dry red. Vars. *hirsutissima* and *angustifolia* shade into each other in some areas. A specimen in the Gray Herbarium from Klickitat Co., Columbia River, Washington, 1906, belongs to the type *O.Hookeri*, having petals 30 mm, a lower leaf 19 cm x 32 mm.

O. Simsiana from Mexico is considered another variety of *O. Hookeri*. The form described as *O. Simsiana* by MAC DOUGAL, VAIL & SHULL (1907) was grown from seeds collected in Mexico City in 1903 by Dr. J.N. ROSE. It had large flowers (petals 40 mm) but a short style, so that the flower was self-pollinated. The catenation is unfortunately unknown[13]. *O.Heribaudi* from Mexico, near Puebla, has very small flowers (petals 12mm) but a long style and is said to resemble *O. sinuata* L. Small flowers in Oenothera are thus by no means confined to the North.

O. franciscana BARTL. is reduced (GATES 1957) to a variety of *O. Hookeri*. As is well known, it generally has a ring of 4 chromosomes, thus showing the beginning of catenation. It was grown at Bethesda, Md., in 1905 (BARTLETT 1914b). The style is long and the leaves moderately crinkled. It has since been used in many crosses with other species. *O. venusta* was grown by BARTLETT from seeds collected at San Bernardino, California, by S.B. PARISH in 1912, and by me from the same collection (GATES 1915, p. 28). It is a much taller plant with larger leaves and flowers. It is given reduced status as a var. of *O. Hookeri*. The leaves are somewhat crinkled, and the bracts, hypanthium and sepals are covered with grey-green viscid hairs. These hairs are absent in var. *grisea* BARTL., presumably through a loss mutation. The differences between var. *franciscana* and the type of *Hookeri* are of minor character (HOEPPENER & RENNER 1929). *Franciscana* differs only in having (1) shorter and broader leaves, (2) less red on the stem. The reciprocal hybrids between them were alike and intermediate, with no clear segregation in F_2.

Small-flowered varieties of many species have now been described. Each must be regarded as a forward step in the evolution of the genus. There is much evidence that numerous dominant mutations must have occurred as successive steps to small flowers (PEASE 1940), but the chromosome catenation produces a form of genetic linkage which makes it nonfeasible to analyze them fully. Probably the same flower-size mutation has occurred independently in many lines of descent, but some will have been transferred by crossing from one line to another. They are measured by petal-length, for generally all parts of the flower are coordinately reduced. The reduction in style is greater, however, thus bringing about a change from open pollination to a condition in which the shedding anthers surround and are in contact with the appressed stigma lobes in the bud. It has been suggested that growth substances are thus transferred to the stigma.

In *O.levigata*, as pointed out by BARTLETT (1914a), the hypanthium elongates faster than the style in the late bud stage, so that the open anthers are pushed up past the stigma, which is thus pollinated in the bud, especially in late flowers, the stigma being finally within

---

[13] *O. Simsiana* SER. is included by MUNZ (1949) in *O. elata* H. B. K. A cytogenetic study of 5 races from the highlands of Central Mexico and Guatemala (STEINER 1951) shows that the chromosome ends differ from those of *O. Hookeri* by two interchanges. This species is thus also at an early stage of ring-formation.

the hypanthium tube. In many other species, in flowers late in the season the style does not decrease in length in harmony with the size-decrease of other parts of the flower. As a result, the stigma may protrude from the bud and may even be air-pollinated before the flower opens. Such late flowers, even if pollinated, have only a very remote chance of ripening seeds before the frosts kill them. But EMERSON (1936a) observed in the Western States that certain insect larvae feed on the anthers in the bud, leaving the style and stigma uninjured. Such a flower could afterwards be cross-pollinated and produce hybrid seeds. It would appear that only moths seeking nectar from the base of the long hypantium would be likely to visit such a flower. Further East, in St. Louis, the very young buds of Oenothera cultures are stung by an insect (GATES 1910b). Such buds fail to develope a hypanthium, the bud cone becoming very thick at the base and frequently developing red pigment.

O. *Reynoldsii* BARTL. from Tennessee and O. *pratincola* BARTL. from Kentucky are distinct species. O. *numismatica* is reduced to a variety of O. *pratincola* (GATES 1957). Similarly, O. *litorea* is treated as a var. of O. *syrticola*. The latter is nearly related to O. *ammophila* FOCKE, described from Bremen, Germany, where it was introduced. It probably came originally from the New England coast. O. *rubescens* from Nantucket Island, Massachusetts, and O. *columbiana* BARTL. ined. (from Washington, D.C.) both have ± red hypanthium but are unrelated. O. *magdalena* from the Magdalen Islands (GATES 1951a) has conspicuous pink hypanthia but is otherwise different. O. *gauroides* HORNEM. is an older species common in Maryland and Virginia. Var. *brevicapsula* from Chevy Chase, Md., differs from O. *gauroides* in its more condensed spikes and shorter fruits (18–20 x 6.5–7 mm). O. *ruderalis* BARTL. is from Chevy Chase, Washington and Baltimore. Its relationships are not clear.

O. *muricata* and O. *parviflora* are both Linnaean species whose original source in eastern North America is unknown. O. *muricata* is characterized by stems bearing red papillae, stem tip bent, small flowers (petals 10-15 mm). Its complexes are *rigens* ♀ • *curvans* ♂. RENNER (1937a) finds the gene R for red midribs in *rigens* identical with the lethal R in O. *biennis*. The *muricata* from Venice with red midribs is classed as var. *rhodoneura*. In the cross O. *muricata* X *rubieflexa* one plant among 84 had a wholly red calyx, which may be a mutation like *rubricalyx*. DAVIS (1914) has carefully listed the differences between O. *biennis* and O. *muricata* as grown in the genetical experiments of himself, DE VRIES and others.

O. *parviflora* has still smaller flowers (petals 8 mm), rosette leaves oblong-lanceolate, strongly denticulate, dark green and shiny, mottled with red; inflorescense dense; buds club-shaped, sepal tips separate, petals cuneate, emarginate. A form which at least resembles it very closely, as shown by the description by J. A. MURRAY,

1776 (quoted in WEIN 1936), was rediscovered by MacDougal (1907) at South Harpswell, Maine, in 1905 and shows certain resemblances to *O. atrovirens*, e.g. in the dark green leaves, shiny, but "mottled with red". This condition I have called liver coloured spots or blotches, as found in *O. Hazelae*.

The type of *O. biennis* L. was shown by BARTLETT (1913) to be the race which has been common in Holland since the 18th century. Its American source can only be conjectured, but it may have been from New England or probably further south[14]. It was shown by RENNER (1917) to be composed of the complexes *rubens* ·*albicans*. The var. *sulfurea* was known to LINNAEUS in Holland when he wrote the "Hortus Cliffortianus" (1937).

DAVIS (1940) obtained from the Cambridge University Botanic Garden a related form which he described as *O. cantabrigiana*, now reduced to a variety of *O. biennis*. It differs from *O. biennis* in having (1) red papillae on the stem, (2) stem becoming red below, (3) midribs light red, (4) broken streaks of red on the sepals, (5) a few papillae on the ovary. This development of anthocyanin in the leaves, stem, ovary and sepals probably results from a single gene difference. From unpublished observations of the present writer, the practically identical form is found in various parts of England and in Wales where it is common at Barmouth. My observations of wild plants at Barmouth in 1933 are as follows: Leaves (including rosette) smooth or slightly crinkled. No red papillae on stems, ovaries or fruits. Somewhat scattered long hairs from small green papillae. Midribs red, stem leaves red at base below. Leaves under a lens show scattered fine pubescence more dense on midribs and main veins. Buds green, showing only short hairs, close set on ovary, sparse on hypanthium, a few on the shining green cone. Petals truncate, 29 x 25 mm, sepal tips 2-4 mm, green, more pubescent, some tipped with red. Anthers touching the lower two-thirds of stigma lobes in the bud. Fruits green, stout.

It will be observed that, like the true *O. biennis*, there are no red papillae on the stem. In my cultures *cantabrigiana* remained quite uniform. Whatever its historical relation to *O. biennis*, it has evidently been long naturalized in Britain. Further investigations will show the changes which are necessary to produce it from *O. biennis* – probably only one gene mutation and a chromosome interchange. DAVIS finds the catenation to be ⑩ + ④ whereas in *O. biennis* it is ⑧ + ⑥. DAVIS also shows that the pollen and egg lethals are reversed in position from those in *O. biennis*. He calls the complexes *rubens* red and *albicans* red. Crosses between the two forms give the viable combinations *rubens·rubens* red and *albicans·albicans* red, whereas

---

[14] From an analysis of five Oenothera races in Western New York State, SLOAT-MAN (1953) finds that their α-complex is *strigosa*-like, the β-complex *biennis*-like.

rubens·rubens and *albicans·albicans* are lethal in *O. biennis*. The relationship to *O. biennis* is so close that *cantabrigiana* should evidently be treated as a variety of *O. biennis*. The indications are that it has probably originated in England since the 17th century.

DAVIS (1926) finds *cantabrigiana* represented by herbarium sheets in Manchester University and from Jersey (1871) as well as at Burnham-On-Sea in North Somerset (1883 and 1886). The latter locality was visited by me in 1938, and the seeds collected produced a uniform culture (9.39) characterized as follows: Stem erect, short, ribbed, green, with red papillae. Leaves lanceolate, finely repandodenticulate, smooth, 21.5 cm x 52 mm, midribs pink. Inflorescence rather lax, comose, apex flat. The ovary, 19 x 4 mm bore long hairs from red papillae; hypanthium 30 x 2,5 mm, stout, yellowish; bud cone 16-27 x 6 mm, bearing colourless papillae; sepal tips, tips 2-5 mm., filiform, terminal, appressed; petals 22-25 x 28 mm, filaments 17 mm, anthers, 8 mm; stigma lobes spreading, 16 mm long (an extraordinary length), 13 mm above hypanthium. This form is not *cantabrigiana*. It belongs to *O. biennis* in the general sense, but it has red papillae and the catenation is ⑭. It appears that *O. muricata* and *O. biennis* have both produced forms with red midribs since their escape from cultivation in Europe. Many specimens of *O. biennis* are cited by DAVIS from British localities since 1800. See also GATES, 1915, p. 17.

Oenothera insignis BARTL.

This species was grown in 1936–39 from several collections of seeds made by Prof. W.P. THOMPSON, of the University of Saskatchewan, and others. Unlike many other Oenotheras, this distinctive form extends over a wide area in Western Canada and as far east as Quebec. In the west it appears to be the only species over large areas. There is evidence that it has recently been extending its range northwards, with the removal of forest and the opening up of N. Saskatchewan to farming. Seven cultures were grown from wild seeds in 1938, and 14 in 1939. The known localities in British Columbia, Alberta, Saskatchewan, Ontario and Quebec, are given in GATES (1957). The species is relatively uniform from Saskatchewan to British Columbia. The cultures in its eastern range in Ontario and Quebec differ in certain particulars, but clearly belong to the same species. The prairie type appears to have spread eastwards. As pointed out elsewhere (GATES 1936, p. 339), *O. insignis* when grown in England loses its (sub-acaulescent) habit of forming flowers in the lowermost leaf axils at or near ground level, and developes a longer stem. It appears to be more plastic than most Oenothera species.

The relation of dimensions to species-description in Oenothera, i.e. the range of variability in size to be allowed, is an ever-present problem to the taxonomist describing a species, and a matter of some difficulty, especially when garden plants are compared with the wild. *O. insignis* is described from cultures from Saskatoon, grown in Regents Park, 1938–39, as follows: Rosette leaves medium to dark green, broad lanceolate, apex acute or acuminate, ca. 9.5 cm long, 55 mm broad, ± crinkled, margin ± wavy, repand-denticulate, midrib white, both surfaces with scattered appressed puberulence. Stem short (26-34 in.), erect, brittle, ± ribbed, green, with or without long hairs from red papillae, but with ascending pubescence and short appressed puberulence; a basal ring of long, nearly prostrate branches, cauline branches many, widely spreading. Leaves 14.5–16.5 cm x 42–48 mm, medium to yellowish green, broad-lanceolate, lower curled or waved, upper smooth; petiole short, midrib broad, generally white, sometimes pink, with ascending pubescence and puberulence, denser below; margin obscurely distantly repand-denticulate with green glands, subentire towards base. Lower bracts 11 cm x 45 mm, waved or deflexed, upper 14–20 cm x 3–6 mm, waved. Inflorescence extremely lax, apex convex or flat, comose. Ovary 14–18 x 3–4 mm, many long ascending hairs from very small green or red papillae, ascending pubescence and dense crisped appressed puberulence. Hypanthium 30–33 cm x 2.5–3 mm, greenish or yellowish with short appressed puberulence. Bud-cone 14–18 x 6–7 mm, squarish, tapering, greenish or ± red at shoulder with spreading long and short hairs (a few from colourless papillae) and appressed puberulence. Sepal tips 2–4 mm, terminal, appressed, green or red at base, slender. Petals 16–19 x 22–23 mm, truncate, opening erect, overlapping. Filaments 10–14 mm, ± arcuate. Anthers 7–9 mm, stigma lobes 7-8 mm., appressed or spreading, 5–7 mm above hypanthium. The alternatives in this description refer to different cultures from Saskatoon, or to size differences of the same culture in different years. Thus in one culture the buds are ± red at the shoulder, in another green; the sepal tips are green in one culture, red at the base in another. In one culture the petals fade to terracotta and the flower remains for some time attached.

After mature consideration of these cultures and comparison with *O. insignis* as originally described by BARTLETT (1914a) and later supplemented by GATES (1936, p. 337), it seemed best to include them all in one widespread species rather than attempt to delineate subspecies or varieties within the complex. This is justified when one finds that the original culture of *O. insignis* (l.c., p. 337) differs from these later ones almost entirely in quantitative characters. Thus the stem height is 75–85 cm instead of 40 cm, but this great difference appears to result from loss of the subacaulescent habit in cultivation. The petals in the 1934 culture were 15 mm long in annual plants, only 10 mm in the biennial, while in the 1938–39 cultures (annual) they were 16–19 mm. The general appearance of the latter cultures can be seen from Figs. 10 (rosette) and 11 (in flower). BARTLETT described the rosette leaves of plants grown at Washington, D.C. from seeds collected near Duluth, Minnesota) as blue green, the stem oblique with bent tip, pale red below, sepal tips slightly infraterminal, the bracts with a long petiole. Notwithstanding all these differences from the present cultures and the fact that the rosette leaves are much wider (55 mm instead of 20–37 mm) we include them in the same species because they agree in (1) frequently having a ring of nearly prostrate basal branches, (2) foliaceous bracts, (3) character of pubescence, (4) white midribs of leaves, and (5) few or no red papillae and these very small when present. Some of the increases in size, as in the length of the ovary, may perhaps be attributed to cultivation. Some Oenothera specialists may prefer to call this a new species.

In their extensive experiments with *Potentilla glandulosa* and

Fig. 10. *O. insignis* rosette. From Saskatoon, Saskatchewan. Cult. 26A.39.

Achillea millefolium in relation to climate (1941), and transplantation experiments with Achillea in California, CLAUSEN, KECK & HIESEY (1948) found that the ecotypes adapted to alpine or lowland climates were fixed in character and did not revert when transported

Fig. 11. *O. insignis* in flower. From Saskatoon. Cult. 26A. 39.

as clones to the other climatic conditions. The fact that these Oenotheras when grown from seed in the English climate lost at once their prairie habit unless they flowered in the first year argues strongly that their prairie habit is not yet fixed and irreversible. This might be because the time that has elapsed since these Oenotheras reached the Canadian prairies is less than the time during which plants in Central California have been subjected to the selective forces of alpine or valley conditions. They still retain the power of reacting in two different ways according to the conditions. TURESSON (1925) showed similarly that adaptive ecotypes such as the prostrate habit of certain species are inherited in some cases and not in others. The ecotype assumes certain modifications as a direct result of environmental impact until such time as a mutation produces the same kind of adaptive type which is inherited. It is probable, however, that the whole story is not as simple as this.

It is also to be noted that a plant of this species from Saskatoon (culture 5.38) was examined cytologically by Dr. T.K. JACOB and found to have ⑫ chromosomes $+ 1_{11}$. This plant from the Middle West of Canada is, with *O. chicaginensis*, an exception to the rule that the smaller-flowered northern forms and especially all those in eastern North America (see list in GATES, 1936, p. 349) have ⑭.

We come now to the cultures of *O. insignis* from other localities. The form from Prince Albert, 100 miles north of Saskatoon (on the northern branch of the Saskatchewan River and at the southern edge of the forest zone) differs as follows: stem $\pm$ fasciated, with pale red stripes, no red papillae visible except minute ones (with hand lens), and very few long hairs; leaves narrow, 10.5 cm x 26 mm, lower bracts 80 x 15 mm, narrow lanceolate; ovary with no red papillae; petals 16 x 15 mm, stigma 2-4 mm above hypanthium. In this culture there was an almost complete absence of papillate hairs on the plants, and the lowest fruits were well up on the stem.

Waskesiu, Sask. is about 80 miles north of Prince Albert and in the heart of the forest. Prof. W. P. THOMPSON collected seeds here from the only plant he saw north of Prince Albert. It must have been recently introduced from further south. In my culture from this source the stigma lobes were 13 mm long, 8 mm above the hypanthium, and the lowest flower was 18 in. from the ground. The culture from Osoyoos Lake, B.C., near the border of Washington State, was almost identical with culture 6.38 from Saskatoon. This culture contained one dwarf plant (21 in. high) with smaller flowers (petals 10 mm) which was probably a haploid. The cultures from Luskville, Hull Co., Ont., were very similar to those from Saskatoon, stems taller (40 in.), stout and ribbed, petals 18 mm, fading brown. The culture (211. 39) from St. Jerome, Quebec, showed great similarity to Fig. 12, but the leaves are less crinkled, the petals 15–16 mm, fading terra cotta.

Cult. 123.38, 64.39 was derived from seeds collected at Rivière Blanche in Eastern Quebec. They were considered very close to the prairie type. But the rosette leaves were yellowish green, 15 cm x 49 mm, the petals 14-17 mm, sepal tips 1-3 mm, green, petals not fading terra cotta. A flowering plant from this culture is shown in Fig. 12. A culture from Adams, Wis. also probably belonged here. These forms all agreed in (1) a ring of $\pm$ ascending basal branches, (2) stem ribbed, brittle, green with few or no red papillae or long hairs, (3) leaves lanceolate, with white midribs and fine puberulence, (4) inflorescence comose, (5) petals 15-20 mm long.

From the measurements given, it will be evident that there is a considerable range of fluctuations in the size of every measurable organ, a universal condition, as every biologist knows; but it is

Fig. 12. *O. insignis*. Rivière Blanche, Quebec. Cult. 64.39.

impossible to determine how wide is the whole range of fluctuating variability. It is this difficulty which makes it impossible to rely solely on measurements in differentiating species of plants. The ranges of variability for many characters often overlap, yet the non-measureable characters may make it clear that we are dealing with distinct species. In this case, the wide difference between the species wild in Duluth, Minn., and as grown in England from seeds collected in Saskatoon and elsewhere, appears to be mainly to reactions to climate.

The late Professor MARIE-VICTORIN and his colleagues, Professors Jacques ROUSSEAU & ROLLAND-GERMAIN, collected seeds as well as specimens of Oenothera from many parts of Quebec

Province. All of my later cultures from this area were grown from seeds they kindly sent me. Their specimens are preserved at the Montreal Botanic Garden Herbarium. The type specimens of the new species and varieties described in my 1936 monograph were destroyed in the war, but the paratypes are at Kew, and other sets of specimens are at the Gray Herbarium, the New York Botanical Garden, and the Montreal Botanic Garden.

Culture Variations

We may now consider the cultures which are more or less aberrant under the various species and varieties. As already stated, the cultures from many new eastern localities fitted exactly into species or varieties already described (see maps). The measurements of aberrant forms, together with those of related species for which measurements have been published, are collected into a large Table but not published here. This Table has been compiled from all the available measurements.

In making these measurements one looked down the row of plants, which are very uniform, and then selected for measurement a plant or a leaf, bud or flower which represented as nearly as possible by inspection the mean or optimum condition for the whole culture. Of course, to have made a series of measurements for each character in each culture and then used the mean would have been more accurate, but would have involved an impossible number of measurements. With care and with many years of practice in observation of these characters, it is believed that the statistical error involved in the method adopted is small and would be measured in fractions of a mm. The diameters of floral organs were generally measured to the nearest half mm, petals and fruits to the nearest mm.

The question of the relation between measurements and phenotypic characters such as pubescence, leaf shape, margin, surface, etc., in the description of Angiosperm species, is a large one. Each has its limitations, but both are required in any accurate description. The limits between hereditary and non-hereditary variations in size measurements can only be set by extensive series of observations in a particular environment of climate and soil, and even then the possible size-changes of particular organs under a different environment will be largely conjectural.

The subject will not be discussed further here, but reference may be made to the work of MATHER (1949) who has made important contributions to the development of biometrics and shows how distinctions may be drawn between inherited and non-inherited quantitative differences. He shows for instance, that polygenes forming continuous series are concerned in species formation, such genes helping to fit the species more closely to its environment.

In a little-known paper (GATES 1932) published in Russia, in a "Festschrift" for VAVILOV, it was shown that the mean petal-length of flowers on the side branches was always 1-2 mm less than on the main stem, regardless of the flower-size. The difference is only visible to the eye in species with small flowers, but measurements showed its existence in the large-flowered species as well. This is clearly a matter of nutrition, the vascular supply being greater in the central stem. Probably the same conditions hold in many other plants. Using meteorological instruments, it was shown that flower size also depends on the temperature at the time of flower-opening. Temperature is thus one cause of the fluctuations in size of flower from day to day. Over 100.000 measurements of petal-length were made in several years, from which the relative stability of flower size in each species or variety became apparent.

In this aspect, genetics comes into close contact with plant physiology. One must remember that all "genes" are ultimately physiological or biochemical in their action. The phenotypic characters produced by the plant in the growth of its various parts are an expression of the relation of the genotype to an ever-varying environment. Different strains of the same species grown simultaneously in the same garden may differ in mean petal-length by only a fraction of a mm. They may not show the same amount of difference if grown together in another season with its different weather conditions, and the meaning of different degrees of "innate" variability in these heterozygous forms is uncertain.

Measurements were made for various strains of *O. novae-scotiae*. The culture (24A. 38, 66.39)[15] from Charny (Levis Co.) Quebec, appears to belong to *O. novae-scotiae*, but with much smaller flowers. This is confirmed by comparing the measurements. The Charny strain differs in being (1) much taller, with (2) much smaller flowers, but (3) longer hypanthium, ovary and stigma lobes, and (4) larger, wider leaves. These are regarded as significant genotypic differences, although the range of fluctuation of these measurements in any genotype is not accurately known and can only be judged on the basis of wide general experience of size variation in the group. In the phenotype and in measurements this Quebec strain otherwise agrees with *O. novae-scotiae*. The strain from St. Hubert (Chambly Co.) south of Montreal agrees closely with that from Charny, but differs in (1) shorter hypanthium and ovary, (2) larger petals, (3) wider leaves, (4) less red on stem and midribs, (5) bracts ± crinkled. In this case, lengths of petals and hypanthium, which usually vary together, have varied in opposite directions. This has happened in certain other strains.

The measurements of cultures from two seed collections from

---

[15] The culture numbers are explained on p. 39.

Prince Edward Island, Tryon[16] and Ellerslie are practically identical. The two series for Ellerslie were made independently by different investigators (GATES & W.R. PHILIPSON) and they are in complete agreement. These measurements also agree with those of *O. novae-scotiae* from the original locality in the Annapolis Valley, except that they have longer stigma lobes (ca. 10 mm instead of 5 mm) and somewhat broader, smooth leaves. The Ellerslie cultures differ from the Tryon strain (and from the type of *O. novae-scotiae*) in having no red papillae on the stem or ovary. We thus have the same species represented in the Annapolis Valley, N.S., in Prince Edward Island and from near Quebec City and Montreal. This striking discontinuity in distribution, unlike that of other species, is not easily accounted for. So many collections of other Oenotheras have been made in the intervening area that it is difficult to believe the distribution is continuous. Perhaps dispersal of seeds by man, e.g., in hay or fodder, may account for this. The highly heterozygous condition of all these Oenotheras will account for the occurrence of single character differences, through a form of crossing-over.

O. novae-scotiae var. *intermedia* differs from the species in several characters which ally it to *O. Hazelae:* (1) narrower, shiny rosette leaves with purple spots, (2) smaller flowers, (3) somewhat bent stem tip, (4) subterminal sepal tips, (5) the very short style. It has probably originated at some time through crosses between strains of these two species; like *O. comosa* below, from which it differs in (1) leaves less narrow, (2) stem tip more bent, (3) presence of hairs from papillae, (4) smaller flowers, (5) short style. The var. *serratifolia,* some 30 miles from the type locality of *O. novae-scotiae,* differs in much smaller flowers with shorter sepal tips, leaves narrower and markedly repand-dentate, and inflorescence more compact. Var. *distantfolia* differs mainly in its narrower stem leaves and long internodes.

O. comosa, from a locality near the type of *O. novae-scotiae,* differs in being taller, with narrower leaves and slightly smaller petals. The non-measureable qualities, darker green leaves with liver-coloured spots, stem tip slightly bent and sepal tips subterminal, are characters of *O. Hazelae.* Long hairs from papillae are nearly absent. These characters make it evident that this species must have been derived at some time from crosses between *O. novae-scotiae* and *O. Hazelae,* both of which occur in this vicinity. It of course breeds true like all the other small-flowered Oenotherae with complete catenation. *O. intermedia* is treated as a var. of *O. novae-scotiae* because it is easily attached to that species, but *O. comosa* appears to contain a more equal mixture of characters of *O. novae-scotiae* and *O. Hazelae.*

---

[16] These seeds were kindly collected by Miss Lilla WRIGHT by arrangement with Mrs. Hazel BELL (see map I).

The 24 cultures of *O. Hazelae* from different parts of Nova Scotia generally showed only minute differences (Table III), as in the number of liver-spots on the leaves, the mean petal-length or the length of the sepal tips. Sometimes small but constant differences occur in cultures from different plants of the same colony. Thus in a colony at Port George (Annapolis Co.) on the Bay of Fundy some plants of *O. Hazelae* var. *parviflora* had smooth and some crinkled leaves. These gave rise to cultures, one of which (122.39) had uniformly smooth leaves with liver spots (Fig. 13), the other (123.39) uniformly crinkled leaves with spots (Fig. 14). The cultures were otherwise alike. Cultures with flowers of intermediate sizes occur from some localities, so that the var. *parviflora* is of purely descrip-

Fig. 13. *O. Hazelae parviflora* with smooth leaves. From Port George (Annapolis Co.), Nova Scotia. Cult. 122.39.

Fig. 14.   *O. Hazelae parviflora*, strain with crinkled leaves. From Port George, N.S. Cult. 123.39.

tive value and does not represent a fixed, distinct and uniform strain. The var. *subterminalis* of *O.Hazelae* belongs to eastern Nova Scotia. In measurements the only marked difference is in the smaller, narrower leaves which are also much crinkled. This crinkling is constant from year to year and has already been mentioned in certain types of var. *parviflora*. Red papillae are more conspicuous on the stem and the sepal tips are very markedly subterminal. The stem tips are bent, but in many strains of *O.Hazelae* the stem is quite erect. Cultures of *O. Hazelae* from Tennycape, Cambridge and Upper Burlington (all in Hants Co., N.S., on the south shore of Minas Basin) were closely similar except for very minor characters (see Table III). The rosette stage is well represented by Fig. 19 in GATES (1936) except that two of them have more liver spots. The difference in frequency of these spots is inherited. See map.

The culture numbers are for 1939. Practically all these cultures were grown also in 1938, and some in earlier years. They bred true

Table III. Different strains of *O. Hazelae* and var. *parviflora* from Nova Scotia

Cult. No.	Locality	Stem	Midleaf (cm × mm)	Leaves	Petals (mm)	Sepal (mm)	tips	Midribs
122	Port George[1]	—	18.5 × 32	smooth no liver spots	13	3	subterm.	red
123	Port George[1]	—	15.5 × 32	crinkled, liver spots	18	4	subterm.	pink
124	E. Chester	—	16.5 × 35		13	2	subterm.	faint pink
125	Tennycape	—	20 × 29	smooth, many liver spots	12	1	subterm.	pale pink
126	Stirling Brook	—	20 × 31–36	smooth, fewer liver spots	13	3	subterm.	pink
127	Upper Burlington	—	18 × 25	smooth, few liver spots	14	2–3	subterm.	white
128	Cambridge N.S.	tip bent	18 × 26	smooth, more liver spots	11–12	4–5	subterm.	pink
129	Lower Five Islands	tip bent	18.5 × 40	str. crink., many liver spots	12–13	3–4	subterm.	pink
130	Advocate Harbour	erect	18 × 38	sl. crinkled, liver spots	13	3	subterm.	pink
131	Brookville	slightly bent	20 × 38	less crink., more liver spots	12	5	subterm.	pink
132	Frayle's Cove	erect, 31 in.	17.5 × 37	± crink., liver spots	14	1–2	subterm.	pink
136	Hubbard's Cove	strongly bent, 39 in	21 × 43	± crink., liver spots	14	3	terminal	pink
137	Meisner	erect, 41 in.	17 × 29	± crink., liver spots	13	2	subterm.	pink
138	Upper Cornwall	erect, 26 in.	18 × 34	smooth, liver spots	9	1	subterm.	pink
139	Robinson's Corner	erect, 32 in.	17 × 37	smooth, liver spots	8	2–3	subterm.	pink
140	Frayle's Cove	erect, 35 in.	16 × 35	± crink., liver spots	14	2	subterm.	pink
141	Northwest Cove	erect, 27 in.	17.5 × 28	smooth, liver spots	8	1–2	subterm.	pink
142	Ingramport	erect, 31–40 in.	18 × 40	shiny smooth, liver spots	14	5	term.	pink
143	Head Margaret's Bay	erect, 33 in.	17 × 32	smooth, liver spots	8	1–2	subterm.	pink
144	Head Margaret's Bay	erect, 28 in.	16 × 39	sl. crin., no liver spots	8	1–2	subterm.	faint pink
145	Hubley	Same as 144, including crinkled bracts and red spots at base of leaves						
146	Lockeport	erect, 35 in.	15 × 34	sl. crink., liver spots	13	2–3	term.	± white
147	Port Mouton	erect, 23–31 in.	15 × 30	smooth, liver spots	8	1	erect	pale pink
148	Middleton	erect, 34 in.	21 × 41	sl. crink., liver spots	12	3	subterm.	pink

[1] Each culture generally contained about 35 plants and was remarkably uniform.

within narrow limits in such features as amount of crinkling and of liver spots on the leaves, degree of bending of stem tip, length of petals and degree of red colour in the midribs. Nos 138, 139, 141, 143, 144 and 147 belonged to var. *parviflora* with petals markedly smaller than the rest, but other cultures showed smaller differences in petal size which are probably constant under uniform conditions. This was a study of the limits of inheritance between strains. To determine those limits more precisely would require a large statistical treatment, but it is clear that the inherited differences between strains can be very small. The stem in *O. Hazelae* is always short, with generally a ring of basal branches.

The three cultures from Port George (Annapolis Co.), East Chester (Lunenburg Co.) and Lower Five Islands (Cumberland Co.) N.S., all belong strictly to *O. Hazelae* but differ constantly in certain minor features. There are marked differences in height, and the Port George plants (62.38, biennial) have shorter buds with shorter sepal tips. The ovary, however, in this strain is longer, the petals of intermediate size, the style very short. The stem is erect, red papillae are rare on the stem, none on the buds, and the sepal tips are subterminal, sprung. The Five Islands strain has a much longer style and somewhat wider leaves. It also has strongly crinkled dark green leaves with many liver spots, and the sepal tips are very subterminal. These are typical examples of the kind of inherited variability which occurs within each species without in any way approaching another species[17].

The culture of *O. Hazelae* from Hubbards Cove (Halifax Co., 107.38, biennial) is taller, with short ovary and sepal tips, rather small flowers and narrow leaves. It will be seen that the petals in these cultures form a graded series in which many small genes must be affecting the petal size. This strain also had strongly bent stem tips, with a tendency to fasciation in the stem. The midribs and stems were less red than in other cultures. This is but one of many instances of constant differences in the degree of anthocyanin production. This form from the south coast of Nova Scotia differs from the strain from Port George on the Bay of Fundy coast only in (1) bent stem tip, (2) stem taller, stouter, and ± fasciated, (3) leaves larger, (4) less red on midribs and stem, (5) flowers slightly smaller. In another culture of *O. Hazelae* (117.39) from Herring Cove (Halifax Co.) measurements were not taken, but a rosette photograph (Fig. 15) shows many liver spots and a different shape of leaf with its broadest point near the tip. The analysis thus shows the presence of innumerable minor strains within the species, without disturbing its integrity as a species.

---

[17] It should be pointed out that many of the differences between strains (including all in petal-size) recorded by measurements were equally visible to the unaided eye in comparing the cultures of growing plants.

Fig. 15. Herring Cove (Halifax Co.), N.S. Cult. 117.39.

On the other hand, the strain from South Maitland (Hants Co.) on the Shubenacadie River cannot be included in *O. Hazelae*. The rosette leaves are much broader, with some crinkling but rare, minute liver-spots. The stems were erect, the stem papillae green, the leaf midribs pink. The broad leaves and erect stem with green papillae are characters of *O. grandifolia*, which is found in adjacent Colchester Co. This appears to be a natural transition between *O. Hazelae* and *O. grandifolia*, ultimately based on the fact of ancestral hybridizations. The S. Maitland strain differs from *O. grandifolia* in (1) taller stems, (2) shorter buds and petals, (3) longer hypanthia, (4) shorter stamens, (5) narrower leaves, and (6) the petals do not fade to pink at base.

In the cross *O. Hazelae* Lockeport x var. *parviflora* Port Mouton the seed parent had petals 14-16 x 18-20 mm, sepal tips 4-4.5 mm, subterminal, stigma lobes 3-6 mm, 1-8 mm above hypanthium. The pollen parent had longer, narrower leaves $\pm$ crinkled, dark green, shining; stem redder, petals 8 x 10 mm, sepal tips 1-2 mm, "sprung"; stigma lobes 1-2 mm long. In the F_1, small flowers were dominant, sepal tips intermediate, long stigma lobes dominant. The cross

O. Lamarckiana X *O. Hazelae* gave ⑭, while *O. grandifolia* X *O. Lamarckiana* had ⑩ $+ 2_{11}$ (PATHAK 1940). In the former cross, the F_1 *(velans·Hazelae)* had an erect, brilliant red stem; a small rosette; stem leaves smooth, narrow, elliptic-lanceolate; hypanthium reddish where exposed; sepals green; sepal tips 6-7 mm *(Lamarckiana)*, erect, subterminal *(Hazelae)*; hypanthium reddish where exposed.

Table IV. Oenothera Catenations

Cult.

1.[18]	St. Eustache	⑭	T. K. JACOB
5.	Saskatoon (to *O. insignis*)	⑫ $+ 1_{11}$	T. K. JACOB
12.	Ile aux Coudres	⑭	T. K. JACOB
13.	*O. Victorini*, England (W. Wittering)	⑩ $+ 4$	T. K. JACOB
18.	Fort Coulange	⑭	T. K. JACOB
101.	*O. Hazelae parviflora*. Port George	⑭	P. N. BHADURI
117.	St. Jerome	⑭	P. N. BHADURI
138.	*O. ammophiloides laurensis*, Port Elgin	⑭	P. N. BHADURI
148.	*O. angustissima* × *O. angust. Quebecensis*	⑧ $+$ ④ $+ 1_{11}$	T. K. JACOB
149.	*O. angust. Quebec.* × *O. angustissima*	⑧ $+ 3_{11}$	T. K. JACOB
150.	*O. Hazelae* × var. *parviflora*	⑩ $+ 2_{11}$	T. K. JACOB
152.	*O. biformiflora* cruc. × *O. angust. Quebec.*	④ $+ 5_{11}$	T. K. JACOB
154.	*O. biform. cruc.* × *O. rubricapitata*		
	(Type I. broad petals)	⑩ $+ 2_{11}$	T. K. JACOB
	(Type II. cruciate)	⑫ $+ 1_{11}$	T. K. JACOB
155.	*O. eriensis* × Long Id., grey sp. (II)	⑫ $+ 1_{11}$	T. K. JACOB
156.	*O. eriensis* × var. *repandodentata*	⑭	T. K. JACOB
157.	var. *repandodentata* × *O. eriensis*	⑩ $+$ ④	T. K. JACOB
159.	*O. albinervis* FARGO × *O. rubricapitata*	⑫ $+ 1_{11}$	T. K. JACOB
160.	*O. albinervis* BARRIE × *O. rubricapitata*		
	(Type I. no red papillae)		
	(Type II. red papillae)	⑧ $+ 3_{11}$	T. K. JACOB
161.	*O. rubricapitata* × *O. albinervis* Barrie, North Dakota.		
	(Type I. (1.3) red pap.)	⑭	T. K. JACOB
	(Type II. (1.5) no red pap.)	⑩ $+$ ④	T. K. JACOB
164.	*O. novae-scotiae grandiflorens* × *O. blandina*	⑧ $+$ ④ $+ 1_{11}$	S. M. SIKKA
	parviflorens × *O. blandina*	⑫ $+ 1_{11}$	S. M. SIKKA
165.	*O. novae-scotiae* (large flowers) × *O. blandina*	⑧ $+$ ④ $+ 1_{11}$	S. M. SIKKA
169.	*O. parva* × *O. blandina*	⑧ $+$ ④ $+ 1_{11}$	S. M. SIKKA
170.	Ile St. Ours × *O. blandina*	⑫ $+ 1_{11}$	S. M. SIKKA
171.	St. Jerome × *O. blandina*	⑫ $+ 1_{11}$	S. M. SIKKA
173.	*O. ammoph.* var. *flecticaulis* × *O. blandina*	⑧ $+$ ④ $+ 1_{11}$	S. M. SIKKA
174.	*O. Hazelae parv.* (Port Mouton) × *O. blandina* Type I	⑭	S. M. SIKKA
175.	*O. Hazelae parv.* (Chester) × *O. blandina*	⑭	S. M. SIKKA

[18] These are all culture numbers of 1938 and the catenations were determined after the list published by GATES & FORD (1938). A few of these confirm determinations already made from other cultures, while differences in three cases of crosses between *O. novae-scotiae* and *O. Lamarckiana* may be ascribed to one of the parental strains being different or having undergone a segmental interchange. Some of these catenations have been published in subsequent papers (JACOB 1940), PATHAK 1940, SIKKA 1940). Many further catenations in species hybrids have been published by CLELAND (1950).

176. *O. pycnocarpa parv.* × *O. blandina*	⑧ + ④ + 1_{11}	S. M. Sikka
179. *O. pycnocarpa cleistogama* × *O. blandina*	⑭	S. M. Sikka
180. *O. paralamarckiana* × *O. Lamk.* Type I	⑫ + 1_{11}	S. M. Sikka
181. *O. Lamk.* × *O. paralamarckinaa* Type II	⑫ + 1_{11}	S. M. Sikka
182. *O. eriensis* var. *niagarensis* × *O. Lamk.*	⑩ + 2_{11}	G. N. Pathak
184. *O. Lamk.* × *O. novae-scotiae* (valley)		
Type I *gaudens*	⑩ + 2_{11}	G. N. Pathak
Type II *velans*	⑩ + 2_{11}	G. N. Pathak
185. *O. Lamk.* × *O. Hazelae* (Middleton)	⑭	G. N. Pathak
186. *O. novae-scotiae* (large flowers) × *O. Lamk.*	⑩ + 2_{11}	G. N. Pathak
(same as 184. Type II)		
187. *O. Lamk.* × *O. novae-scotiae* (large flowers)	⑧ + ④ + 1_{11}	G. N. Pathak
(same as 184. Type I)		
188. *O. grandifolia* × *O. Lamarckiana*	⑩ + 2_{11}	G. N. Pathak
189. *O. Lamk.* × *O. grandifolia* Type I	⑫ + 1_{11}	G. N. Pathak
Type II	⑥ + ④ + 2_{11}	G. N. Pathak
194. *O. Lamk.* × *O. grandifolia* Type I	⑫ + 1_{11}	G. N. Pathak
Type II	⑥ + ④ + 2_{11}	G. N. Pathak
195. *O. sackvillensis* × *O. Lamk.*	⑧ + ⑧	G. N. Pathak
197. *O. sackvillensis* × *O. Lamk.* Type I	⑧ + ④ + 1_{11}	G. N. Pathak
Type II	⑧ + ④ + 1_{11}	G. N. Pathak
198. *O. ammoph. laurensis* × *O. Lamk.* Type I	⑭	G. N. Pathak
Type II	⑭	G. N. Pathak
Type III	⑭	G. N. Pathak
199. *O. Lamk.* × *O. ammoph. laurensis*	⑫ + 1_{11}	G. N. Pathak
200. *O. ammoph. laurensis* × *O. Lamk.*	⑭	G. N. Pathak
201. *O. parva* (Bic) × *O. Lamk.*		
Type I *(velans)*	⑧ + 3_{11}	G. N. Pathak
Type II *(gaudens)*	⑫ + 1_{11}	G. N. Pathak
202. *O. Lamk. (velans)* × *O. parva (L'Islet)*	⑥ + ④ + 2_{11}	G. N. Pathak
203. *O. parva* (Trois Pistoles) × *O. Lamk.*	⑧ + 3_{11}	G. N. Pathak
204. *O. biformiflora cruc.* × *O. Lamk.* Type I	⑧ + ④ + 1_{11}	G. N. Pathak
Type II	⑧ + ④ + 1_{11}	G. N. Pathak
205. *O. Lamk. (gaudens)* × *O. biformiflora cruc.*	⑧ + ④ + 1_{11}	G. N. Pathak
206. *O. biformiflora* (broad) × *O. Lamk.*		
Type I (pink midribs)		G. N. Pathak
5 plants Type II (white midribs)	⑩ + 2_{11}	G. N. Pathak
207. *O. Lamk. (gaudens)* × *O. biformiflora*	⑧ + ④ + 1_{11}	G. N. Pathak
209. *O. St. Jerome* × *O. Lamk.*	⑩ + 2_{11}	G. N. Pathak
312. *O. deflexa* × *O. Lamk.*	⑥ + ④ + 2_{11}	G. N. Pathak
214. *O. deflexa bracteata* × *O. Lamk.*	⑩ + 2_{11}	G. N. Pathak
215. *O. deflexa* (Yawkey) × *O. Lamk.*	⑥ + ④ + 2_{11}	G. N. Pathak
222. *O. pycnocarpa parv.* (Hamilton, N.Y.) × *O. Lamk.*	⑩ + 2_{11}	G. N. Pathak
223. *O. pycnocarpa cleistogama* (Clinton, N.Y.) × *O. Lamk.* Type I	⑩ + ④	G. N. Pathak
224. *O. pycnocarpa cleistogama* × *O. Lamk.*		
Type I	⑩ + ④	G. N. Pathak
Type II	⑧ + ④ + 1_{11}	G. N. Pathak
226. *O. pycnocarpa parv.* (Georgetown, N.Y.) × *O. Lamk.*	⑩ + 2_{11}	S. M. Sikka
227. *O. blandina* × *O. paralamk.* Type	④ + ④ + 3_{11}	S. M. Sikka
IV. 3.mut.[19]7_{11}		S. M. Sikka
228. *O. blandina* × *O. novae-scotiae*		
1.1 Type I. *parviflorens*	⑫ + 1_{11}	S. M. Sikka
11.7 Type II. *grandiflorens*	⑫ + 1_{11}	S. M. Sikka

[19] Narrow leaves, midleaf 13,5 cm × 20 mm.

Cultures 113.38, 114.39 from East River Point (Lunenburg Co.) and 114A.39 from Blockhouse (Lunenburg Co.) N.S., agree with *O. grandifolia* in having large, broad leaves, but the former hla smaller and the latter larger flowers. The East River strain nearys agrees with *O. grandifolia* also in ovary and hypanthium dimensions, as well as in having green papillae on stem and ovary and erect stems, but differs in (1) much smaller flowers with (2) shorter filaments, anthers and style, (3) pink midribs. Among only 15 plants in the 113.38 culture was a mutation, presumably trisomic, with extremely narrow leaves (midleaf 15 cm. x 12 mm) and petals only 6 mm long. Similar narrow-leaved trisomic mutations have been described (GATES 1936) in *O. ammophiloides* var. *flecticaulis* and var. *laurensis*. The Blockhouse strain also has large leaves and erect stems with green papillae on bud cone and ovary, like *O. grandifolia*, but differs in having (1) much larger flowers with (2) very long sepal tips[20], (3) very long filaments and stigma lobes, (4) pink midribs, (5) red papillae on stem. These two forms thus show important resemblances to *O. grandifolia*, but otherwise differ very much from each other. They may be regarded as variants from the *O. grandifolia* type, produced partly by crossing. Another culture, from Quinpool Road, Halifax, N.S., is nearest to the Blockhouse strain as regards floral measurements, but with somewhat narrower leaves. As in *O. grandifolia*, the stem is erect, bearing green papillae, the leaves broad with white midrib, but the flowers have a strong perfume.

Culture 242.39 was derived from seeds collected by Mrs. WINTHROP BELL on the ridge back of the head of St. Margaret's Bay (Lunenburg Co.) N.S. It agrees closely with *O. ammophiloides* in most measurements, thus extending the range of the species far down the coast from Guysborough. The stem tip was crozier-shaped, the bud cone and hypanthium tinged with red on the side exposed to light. These features and the rather narrow leaves are characters of *O. ammophiloides*, but this strain differed in having (1) bright red stems, (2) buds much less hairy with fewer red papillae, (3) smaller petals. *O. flecticaulis*, originally described from the Lahave River mouth (GATES 1936, p. 269), now reduced to a variety of *O. ammophiloides*, agrees with the latter in (1) the strongly bent stems, (2) increased anthocyanin production on the buds where exposed to light, (3) narrow leaves; but differs in (1) leaves still narrower, (2) flowers much smaller (petals 9-12 mm instead of 18 mm). The Mill Cove (Lunenburg Co.), N.S. culture (114.38 and 115.39) agrees in essentials with var. *flecticaulis* but differs in (1) papillae on stem green, (2) leaves smooth, margin not wavy, (3) no red papillae on buds, (4) midribs white, (5) petals larger, (6) sepal tips shorter.

The var. *laurensis* of *O. ammophiloides* covers a wide territory as a

---

[20] The sepal tips are green, terminal, appressed and slender.

coastal form, extending up the Gulf of St. Lawrence from the border of Nova Scotia and New Brunswick to the Matapedia area, in southern Quebec, and around the southern coast of Gaspé to New Carlisle (Bonaventure Co.) Quebec. Several seed collections were made by me in 1935 in an area within ten miles of Matapedia, where numerous colonies of Oenothera grew, some of them quite luxuriant. The detailed descriptions and the measurements show that they all belonged to the same type, which is included in *O. ammophiloides* var. *laurensis* (Table V). One colony was very tall, another very short. The sepal tips were uniformly 5 mm except in one culture. The hypanthium ranged from 25 to 37 mm. The ovary exhibited a wider range, being twice as long in some cultures as in others. The wide variation in petal length was nearly parallel to that in the ovary, but this is not the case in other species. The measurements of these seven cultures (Table V) are, in general, correlated with flower-size. It will be seen that leaf-width in these seven strains also varies in general with flower-size, as though differences in growth energy of the different biotypes were involved.

Table. V. Variations in the Matapedia cultures

	Stem	Rosette leaf mid-ribs	Stem	Papillae on Ovary	Sepals	Sepal tips
Matapedia 1	crozier, becoming erect	pink	red	reddish	where exposed	subterm., ± red
Matapedia 2	tip bent	red	red	few red	red	red on inner face
Matapedia 3	erect	white	green	green		
Matapedia 4	erect	pink	green	green		terminal, green
Matapedia 5	erect	pink	scatt. red	red	colorless	terminal, green
Matapedia 6	tip bent	pink	red	scatt. red	scatt. red	subterm., green, red on innerface
Matapedia 7	tip bent		red	red	red	subterm., green tipped with red

When the Matapedia strains (Table V) are compared with the type of var. *laurensis* M.1. is nearest in foliage but much taller. The stem was crozier-shaped, later becoming nearly erect; sepal tips strongly subterminal, sprung, ± red, and hairy. All these cultures had pink to red midribs except M. 3 in which they were white like the type of var. *laurensis*. In M. 1, 2, 5, and 6 the stem papillae were red, in M. 3 and 4 green. M. 2 differs markedly from var. *laurensis* in (1) smaller flower measurements, (2) red midribs, (3) narrower leaves. The stem was bent horizontally. M. 3 agrees closely with M. 2 in measurements, except in taller stems and broader leaves. It had an erect stem, white midrib and green papillae on stem and ovary. M. 4 had an erect, very brittle stem, leaves slightly crinkled, with pink midribs. The papillae on stem and ovary were green, sepal tips green, slender, terminal, appressed, the flowers much larger, stem tip bent. In culture M. 5 the stem was erect, midribs pink, and red papillae on stem and ovary. The flowers were smaller than in M. 4, the leaves somewhat narrower, sepal tips like those of M. 4 but shorter, flowers smaller and with a delicate odour, style shorter. In culture M. 6 the stem was strongly bent, tough, green with red papillae, leaves slightly crinkled, midribs pink, conspicuously pink below at base, scattered red papillae on ovary, hypanthium and bud cone. It differs constantly from M. 5 in (1) bent stem tip, (2) petiole and midribs red below, (3) midribs stronger pink, (4) sepal tips subterminal, red on inner face. The flowers, like M. 5, have a delicate odour. The culture (7) from New Carlisle (Bonaventure Co.) Quebec, differs from M. 6 in the narrower leaves, shorter hypanthium and longer ovary.

All these cultures show in varying degrees a marked feature of var. *laurensis*, i.e. sensitiveness to light, producing more anthocyanin on stems and buds where exposed. So far as they can be tabulated, the main non-measureable differences in these seven cultures are classified in Table V.

Such a distribution of characters in colonies of a single self-pollinating species in a particular area looks superficially like Mendelian segregation and recombination, but the mechanism behind it is more complex, presumably depending on rare crosses and occasional crossing-over between the catenated chromosomes, and still rarer segmental interchange between non-homologous chromosomes. It is possible, but hardly likely, that all the genes involved are in the pairing-ends of the chromosomes.

A culture from Mont St. Pierre (Gaspé Co.) Quebec (132.38 and 83.39) appears to be transitional to var. *parva*. It has small, smooth, narrow, greyish-green leaves with white midribs and without liver spots, like *parva*. The stem tip is bent and there are many red papillae on the stem, especially on sepals and fruits where exposed to light. These light-sensitive papillae are found in var. *parva* but are especially characteristic of var. *laurensis*. The fruits are longer than the mean of either variety.

Culture 21.39 was grown from seeds collected at Cape Gaspé by Mrs. B.W. TAYLOR and transmitted by Professor V.C. WYNNE-EDWARDS. It was uniform and differed from *O. ammophiloides* var. *laurensis* of Port Elgin, N.B. in (1) much smaller flowers (petals 12 mm, (2) fewer red papillae on buds, (3) narrower leaves (midleaf 14 cm x 18-19 mm). The New Carlisle culture 23.39, from seeds collected by Dr. H.F. LEWIS and transmitted by Professor WYNNE-

EDWARDS, contained two distinct types (probably from different plants) having respectively 17 with broader (16-25 x 34-48 mm) and 5 with narrower midleaf (16.5-19.5 cm x 23-29 mm), 5 others being intermediate and unclassifiable. The petal size was uniform, 16-18 mm. The amount of red on the buds varied markedly but could not be scored. This culture represents *O. ammophiloides*, differing only in minor features from var. *laurensis*.

Culture 23.38 and 80.39 from St. Joachim (Montmorency Co.), on the north bank of the St. Lawrence, discloses a form which agrees with neither of the previous varieties though it has characters of both. It is described as *O. ammophiloides* var. *angustifolia*.

Fig. 16. *O. ammophiloides* var. *angustifolia*, St. Joachim (Montmorency Co.), Quebec. Cult. 80.39, rosette.

Fig. 17. *O. ammophiloides* var. *angustifolia*. St. Joachim, Quebec. Culture 80.39, in flower.

The stem bears red papillae and the tip is bent horizontal (Fig. 17). The leaves are narrow (Fig. 16), with white midribs, as in var. *parva*, but not grey-green. The margin is subentire above, distantly repand-dentate below. The flowers are very small (petals 10 x 11 mm) but the hypanthium is long (30 mm). There are no long hairs or papillae on the ovary, hypanthium or bud cone. The sepal tips are very short, green, terminal, appressed. Culture 79.39, from Ile aux Coudres (Charlevoix Co.) on the St. Lawrence, is closely similar to the St. Joachim culture while culture 22.39 from Mont St. Pierre is typical var. *parva*.

The culture (24.39) from Dolbeau (Saguenay Co.), Quebec (Fig. 18) agrees with that from St. Joachim (Fig. 16) in essentials, except that the rosette leaves are ± crinkled and the flowers reach the extremity of small size (petals 7-8 mm). Rosette leaves dull green, narrow lanceolate, gradually narrowing below to a broad unmargined petiole, crinkled and strongly wavy, midrib white, subentire above to repand-dentate below, subglabrous, fading in red blotches. Stem green, with many large red papillae, the tip bent, inflorescence rather lax. The ovary and hypanthium bear scattered red papillae.

Fig. 18.   Dolbeau (Saguenay Co.), Quebec. Culture 24.39.

The petals open out flat, leaving spaces between them. The sepal tips are green, stout, terminal, appressed, with many red papillae and some diffuse red. The stigma lobes remained appressed instead of opening out flat, but this is not always a reliable differential character. In some forms they may be either appressed or open in anthesis. The cross, *O. ammophiloides* var. *laurensis* X *O. Lamarckiana* (Table IV) gave three types in F_1, all with ⑭, while the reciprocal gave ⑫ $+ 1_{11}$ (PATHAK 1940). On the other hand, var. *parva* X *Lamarckiana* produced two types, one having ⑧ $+ 3_{11}$, the other ⑫ $+ 1_{11}$, while the reciprocal had ⑥ $+$ ④ $+$ ④.

The locality on Lake St. John is the most northerly point from which Oenothera cultures have been made. The seeds were collected by Messrs. MARIE-VICTORIN and ROLLAND-GERMAIN. It is interesting that the smallest flowers should have come from this latitude ($48°$ 36' N.), but the relationship between flowersize and latitude only holds in a very general sense. For example, *O. Victorini* var. *parviflora* from south of the St. Lawrence has flowers almost equally small. Probably some of the forms with intermediate flower-size, such as the "large-flowered" strain of *O. novae-scotiae*, have been derived secondarily from smallflowered forms through loss of the dominant genes for smaller flowers.

In these Oenotheras with small flowers, which do not depend on insect visits for fertility, the petals could be completely lost without any disadvantage. That they are necessary for protection in the bud is very doubtful. It is conceivable that future evolution may produce a "new genus" of apetalous plants. This would be the logical end of the present tendency. It is probably the method by which some apetalous species have developed in other families of Angiosperms in the past. Mut. *apetala* from *O. suaveolens*, found by DE VRIES, has already been mentioned (p. 24).

A culture (124.38, 104.39) from Wolfe's Cove, Quebec, was from seeds collected by the writer from a larger colony under the cliffs. It belongs to *O. biformiflora*, a species in which the petals may be either broad or cruciate (GATES 1936, p. 303), the two forms often occurring intermingled in the same colony. The flower measurements of this strain agree closely with those of the type culture from the south bank of the St. Lawrence but (1) the stem leaves are wider, (2) the stems brittle, (3) the papillae on stem and ovary are green, (4) the sepal tips, style and stigma lobes are longer, (5) the leaves have a few liver spots. The strain agrees with *O. biformiflora* in having short, green, slender, terminal sepal tips, but the petals in this large colony were all broad.

It will be apparent that many of the differences recorded in these cultures will depend upon single (or in some cases multiple) genes. But strains cannot be safely analyzed in these terms until actual crosses have shown the method of inheritance. As regards catenation,

the cross $O.$ *biformiflora cruciata* X $O.$ *angustissima* var. *quebecensis*
gave an F_1 with ④ $+ 5_{11}$ (JACOB 1940), indicating a closer similarity
of the chromosome ends than would be expected from these very
different phenotypes. A pot full of sublethal yellow seedlings was
produced, only two of which survived as rosettes and came to
flower. The rosette leaves had large blotches of yellow and green.
The stem leaves were narrow *(angustissima)*, stem green and midribs mostly white *(biformiflora)*, flowers cruciate, petals 7-8 x 1-2
mm. $O.$ *biformiflora cruciata* X $O.$ *rubricapitata* from N. Dakota gave
⑩ in the broad-petalled F_1 type and ⑫ in the cruciate type (JACOB
1940).

The cultures from St. Eustache (Two Mountains Co.) and Ile
d'Orleans are closely similar in measureable characters, and they
belong with $O.$ *biformiflora*. The St. Eustache form differs chiefly
in the long sepal tips, and a unique feature is that one of the four
sepal tips is a l w a y s m u c h s h o r t e r than the other three. It is
very difficult to picture the kind of germinal change which could
produce such a departure from symmetry. In this connection it is
worth pointing out that (GATES 1930) $O.$ *ammophila* FOCKE[21] from
Bremen and Helgoland sometimes produces pentamerous flowers,
and in $O.$ *novae-scotiae* they may be as high as 15-25% of the flowers.
In $O.$ *ammophila* X $O.$ *novae-scotiae* F_1 all the early flowers were
completely pentamerous, the later ones 4-parted. As the late flowers
also decrease in size, the change from pentamery to tetramery is
regarded as a sign of decreasing growth energy in the plant. In the
Myrtales, to which the Onagraceae belong, floral tetramery prevails,
but, for example, in some species of Eucalyptus, 4- and 5-parted fruits
may be equally common on the same tree. Something in the meriste-
matic apex of the floral primordium determines whether it will
produce a 4- or 5-parted flower. Possibly pentamery is derived from
a larger floral apex, but this is only conjecture.

Although agreeing so nearly in measurements, the St. Eustache
and Ile d'Orleans strains differ in certain other respects. The latter
has short sepal tips like the type of $O.$ *biformiflora*. They agree in
having erect stems, white midribs and no red papillae on stem or
ovary, but the bud cones of the Ile d'Orleans strain contract
abruptly to the short sepal tips, which are subterminal, erect and
spreading (not terminal and appressed). Thus the differences are
practically confined to the sepal tips, although the Ile d'Orleans
strain has a somewhat longer style.

The seed collections from these two localities contained only
broad-petalled flowers, but the possibility of cruciate petals, as in
$O.$ *biformiflora*, is not excluded. The inheritance of the cruciate type

---

[21] BAERECKE (1944) has identified the chromosome ends in $O.$ *ammophila*, $O.*
parviflora, and other species. See also RENNER (1943b und 1945).

in Oenothera is discussed in GATES (1936, p. 308). BARTLETT (1914b) has cited and described various cruciate species and varieties of independent origin, including *O. atrovirens* SH. & BARTL., *O. venosa* SH. & BARTL. and *O. stenomeres* BARTL. As pointed out elsewhere (GATES 1936, p. 312), *O. biformiflora* is probably nearly related to *O. cruciata* NUTT. from Massachusetts. *O. Robinsonii* (BARTLETT 1915d) from Jaffrey, New Hampshire which was cultivated by DE VRIES, differs from *O. venosa* from Hudson Falls, N.Y., only in (1) smaller size, (2) leaves more dentate, (3) narrower bracts, (4) shorter sepal tips, characters which are mostly obscure in herbarium material. *O. stenomeres* BARTL. from Maryland is allied to the broad-petalled *O. gauroides* HORNEM., and *O. stenopetala* BICKN. from the island of Nantucket is similarly related to *O. Oakesiana* S. WATS. from Long Island. Following FERNALD, it is treated here as a var. of *O. cruciata*, but it might better perhaps be considered a var. or derivative of *O. Oakesiana*. *O. biennis* var. *cruciata* shows a separate origin of the same condition. It was found in Holland in 1900.

It is pointed out elsewhere (GATES 1936, p. 309ff) that OEHLKERS found several *cr. (cruciata)* alleles in crosses of various species. The inheritance of petal l e n g t h follows the same general rules, many petals of irregular shape being produced, both in cruciate x normal and in large x small flower. This is another example of the fact that petal size and shape (see below) are partly determined by the cyto-plasm. OEHLKERS (1930, 1938) concluded that a series of probably multiple alleles, Cr_1, Cr_2, Cr_3, Cr_4 . . . cr_1, cr_2, cr_4, cr_5 were involved in crosses of broad-petalled with cruciate varieties. He (1935) found labile conditions in *Epilobium hirsutum cruciatum*, but STOMPS (1913), in a preliminary study of *typicum* X *cruciatum* found the normal dominant, with a sharp 3 : 1 segregation in F_2.

In *O. Lamarckiana* X *cruciata* the cruciate character was domi-nant to normal in some cases, sharp segregation being obtained only with *O. Lamarckiana* & *O. Hookeri*. In *O. biennis* the flowers on one branch were frequently more cruciate than on another, and this difference remained in the next generation. This behaviour was the same as for petal-l e n g t h in *O. biennis* hybrids (GATES 1923). RENNER (1937a, 1939) has also investigated the *cruciata* character in *O. atrovirens* and shown that *Cr cr* can change somatically to *cr cr* or intermediate flowers. He suggests that the gene becomes labile, and in a further analysis (RENNER 1942c, RENNER & SENSEN-HAUER 1942) finds that in hybrids with *O. atrovirens cr* can mutate back to the stable *Cr*. For further studies of *cr* in Oenothera and Epilobium see RENNER and others (1952).

The Three Rivers, Quebec, strain (3.38 and 90.39) belongs to *O. Victorini* var. *parviflora*. It agrees in the small flowers (but slightly larger), lax inflorescence, short sepal tips, erect stems, with absence of red papillae from stem, ovary and sepals. But the style is longer and the leaves larger. The culture (17.39) from Mistassini, Que., also belongs with var. *parviflora*, but with taller stem and somewhat larger flowers, making its measurements almost identical with those of the Three Rivers strain, except that its leaves are narrower, the hypanthium shorter and the stigma lobes longer. The erect stem of this northern form also bears very rare red papillae and some diffuse red, and the subentire stem-leaves generally have a white midrib, are ± crinkled along the midrib, and are strongly waved or curled. The strain (47.38 and 91.39) from Quyon (Pontiac Co.) on the Ottawa River, belongs typically to *O. Victorini*, especially in its bud characters, but in most flower measurements is between the species and the var. *parviflora*. The style, however, is very short. This culture nearly agrees with the Mistassini colony in (1) lax inflorescence, (2) scattered red papillae on stem but none on buds, (3) leaves ± crinkled and wavy, but the midribs are pink.

The main characters of *O. Victorini* are recognized over a wide area from the St. Lawrence below Quebec to Jordan (Lincoln Co.) and Colchester (Essex Co.) Ontario and Piseco (Fig. 19) in northern New York.

An introduced strain of *O. Victorini* occurs spontaneously at West Wittering, Sussex, England (GATES 1936, p. 320). The var. *parviflora* is not in any sense a unity, but the term applies to small-flowered forms of the species occurring in various parts of its range or in outlying areas such as Mistassini. The cultures from Jordan and Vineland at the western end of Lake Ontario agree in all measurements except that the latter has smaller flowers. The culture from Piseco (Fig. 19) in the Adirondacks agrees with that from Vineland except in the longer hypanthia and ovaries. All three agree with the characters of O. *Victorini* in all essentials such as (1) the erect stems with green papillae, (2) the yellow-green buds without red papillae, (3) the slender, terminal, appressed sepal tips, and (4) the rather broad, smooth leaves. A culture from Montreal, shows a typical rosette except that the leaves are slightly crinkled and have a few liver spots. These three strains all differ from *O. Victorini*, however, in having a very lax inflorescence. In the Piseco strain (Fig. 19) the colourless papillae on the stem are elongated. I have seen this character in the conspicuous red papillae on the buds of a strain of *O. Hookeri* from Lake Merced, California. In some buds they were so large and numerous as to form a mass of red granulations on the sepals, and some of the plants growing wild in a fertile soil by the Lake were 9 feet high.

Seven cultures were grown in 1939 from seeds collected by Gilbert GARLICK in the vicinity of Vineland and Jordan (Lincoln Co.) at the western end of Lake Ontario. They all belong to *O. Victorini*, but each culture shows certain differences from the others, as recorded in Table VI.

Fig. 19.　*O. Victorini* var. *parviflora*. From Piseco (Hamilton Co.), New York. Culture 8.39.

As each culture was uniform, these differences are inherited. About 25 plants were grown in each culture, and the cultures were side by side under uniform conditions. There were therefore marked inherited differences in height and in petal length as well as in earlier or later flowering and the other feactures recorded in Table VI. An appearance of ordinary genetic segregation is thus simulated. Each plant in the population is doubtless heterozygous for many genetic differences. The degree of genetic independence of the various characters is perhaps surprising, considering that all these plants have a complete ring of 14 chromosomes.

Table VI. Variations of *O. Victorini* **in one local district**

Cult. No.	Height	Petals	Midribs	Rosette leaves	Liver spots	Leaf surface	Flo-wering
	in.	mm					
1	–	14	pink	jagged below	none	smooth	late
2	–	9	white	margin toothed and waved	rare	rough	–
3	81	13	white	no jags, teeth obscure	none	smooth	late
4	75	12	pale pink	–	present		late
5	50	17	–	like No. 2	rare	slightly dar-ker than No. 1	early
6	75	14–15	pale pink	–	more than No. 4	–	later
7	62	15	white	same as 3	none	smooth, slight-ly darker than No. 1	early

A culture (35.38, 49.39) from Lake Aylmer, Wolfe Co., Quebec, has buds which resemble *O. Victorini* var. *parviflora*, but the foliage is more like that of *O. Lamarckiana* except that the leaves are smaller, less crinkled, and with rare liver spots. The stem is erect, bearing few or no red papillae; rosette leaves broadly lanceolate, slightly or finely crinkled; the basal branches become erect and begin to flower before the stem; the inflorescence is very long; and the floral measurements are generally somewhat larger than in var. *parviflora*. Unlike *O. Victorini*, there are very scattered minute red papillae on the ovary. The sepal tips, as in var. *parviflora*, are very short, green, filiform, terminal, ± tipped with red. The bud charac-ters clearly relate this strain to *O. Victorini*, but whence the leaf characters arise is not clear. Closely related to the Lake Aylmer culture is one from Longueil, south of Montreal, collected by the late Professor MARIE-VICTORIN in sands along the railway. The floral measurements were very similar but the leaves were narrower. In both forms the stem was erect with a ring of suberect basal branches, but the stem papillae of Lake Aylmer were colourless while those of Longueil were faintly pink. The midribs, white in the former, were pink only at the base of leaf in the latter. In both, the ovary bore minute pink papillae, the sepals having no papillae in the former and colourless papillae in the latter. The sepal tips of Longueil differed from those of Lake Aylmer (described above) in being sub-terminal, ± erect prongs, green tipped with red.

The culture from Laniel (Temiskaming Co.) on Lake Kipawa, is closely similar to *O. Victorini* in floral measurements but has narrower petals and broader leaves. The other bud resemblances are close and there are small red papillae on the ovary. The dense inflorescence with relatively few flowers is another point of agreement, but the erect stem has scattered red papillae and the rosette leaves show liver spots.

Of a culture from Thurso (Papineau Co.) on the Ottawa River, surviving notes only indicate rather broad, crinkled rosette leaves (26.5 cm x 60 mm), light green, midrib pink, margin ± wavy. Culture 18.38[22], 77.39, from Fort Coulange (Pontiac Co.) on the Ottawa, differs in measurements from the Laniel culture in having smaller flowers (petals 16 mm) and broader leaves. It agrees with *O. Victorini* in the erect stem with no red papillae, but the very broad rosette leaves are somewhat crinkled, have a few liver spots and white midribs, and the inflorescence is extremely lax. The buds are green, becoming yellow, as in *O. Victorini*, and the sepal tips are green, terminal, spreading at the ends. The most striking difference from *O. Victorini* is that many of the flowers are cleistogamic.[23] With all these differences it may be regarded as a distinct microspecies, but it falls within the range of *O. Victorini* var. *undulata*. In 1939 it was observed that in dry, fine weather all the flowers opened, but in dull, cool weather they were all cleistogamic. STOMPES (1957) describes *O. biennis* var. *hemicleistogama* as a recent cleistogamic mutation from *O. biennis*.

O. pycnocarpa var. *cleistogama* was described from Clinton, N.Y., from seeds collected by Dr. G.L. STEBBINS, JR. (GATES 1936, p. 252). Many observations of cleistogamy were made in these cultures (GATES 1936, p. 250). Some strains had a higher frequency of cleistogamic flowers than others, and there was a definite relation of anthesis to the weather conditions. When the force of petal growth, which normally opens the flower, is sufficiently decreased, the flower will fail to open. In occasional flowers the sepals will be pushed apart only at the base before the growth force fails. Unopened buds often remain for long attached to the plant, especially in dry weather. These sometimes completely fail to produce seeds, while those which become detached after normal self-pollination in the bud produce a full quota of seeds. *O. cleistantha* SHULL & BARTL. was described from Huntingdon, Long Island, N.Y. It shows no relationship with the Fort Coulonge strain. The petals are cruciate and the only similarities in floral measurements are in the bud-cone and ovary. *O. stenomeres* BARTL. from Chevy Chase and Bethesda, Maryland,

---

[22] The catenation of this culture (JACOB 1940) was found to be ⑭.

[23] It is worth pointing out that cleistogamy has been observed as a mutation in hybrids of *Gossypium* (*Indian Cotton Gr. Rev.* 3 : 69, 1949). In *Ruellia strepens* (Acanthaceae) it occurs as var. *cleistantha*, apparently as a mutation (GRANT 1955).

is also both cruciate and (in some flowers) cleistogamic. *O. cleistogama* LÉVEILLÉ from California belongs to the nearly related genus Bois-duvalia. *B. cleistogama* CURRAN fails to open its early flowers. One must suppose that cleistanthy, like cruciate petals, has arisen inde-pendently in different stocks through parallel mutations, but there may be a closer relation between these two conditions, both of which indicate a lack of growing power in the petals.

The strains from Thurso, Fitzroy (Carleton Co.) and Gananogue (Leeds Co.) Ont. (Fig. 20), while related to *O. Victorini*, can not be included within it with propriety, but they can be attached to var. *undulata*. They are obviously similar in having very large rosette

Fig. 20.   Gananogue (Leeds Co.), Ontario. Culture 231.39.

leaves $\pm$ pinnatifid at base and with white midribs, but the latter is more crinkled and the former has liver spots. Five cultures were grown from the Fitzroy area from seeds collected by Prof. Roy Fraser in 1934. They were all very large, luxuriant plants, reaching over 8 ft. in height. The liver spots were absent from one of these cultures which nevertheless had pale pink midribs, showing that the presence of liver spots is independent of midrib colour. Dr. D.G. Catcheside found ⑭ in three different plants from these cultures. The strains from Thurso, Fitzroy and Gananogue all represent the same rather distinct type (see map 2).

O. Victorini as here interpreted with its many varieties, is found in a wide area from the St. Lawrence below Quebec to Lake St. John, also far up the Ottawa River and west to Toronto, the western end of Lake Ontario and into northern New York. *O. levigata*, on the other hand, shows mainly a north-south distribution. Originally described by Bartlett from West Virginia, together with *O. scitula* which was reduced to variety rank (Gates 1957), forms related to it from the St. Lawrence River were described as vars. *similis* and *rubripunctata*. The main characters belonging to this species thus have a great north-south extension which would correspond with its post-glacial dispersion.

O. argillicola. The beautiful *O. argillicola* with large flowers (petals 30-40 mm), glabrous hypanthium and sepals and very narrow leaves, came from the same locality in W. Virginia, where it had a ring of 4 chromosomes and 5 free pairs, while Cleland found a strain at Huntingdon, Pa., with 7 free pairs. This confirms the primitive character of the large-flowered species and the later extension of small-flowered derivatives northwards into Canada. One is also justified in concluding that *O. argillicola* is not a derivative of *O. Hookeri* but represents an equally primitive species which perhaps retreated into Florida during the ice advance, later moving slowly northwards.

A culture (9.38, 78.39) from Les Escoumains, Saguenay Co., Que. from seeds collected by Miss Marcella Gavreau, which was strongly biennial and only flowered in 1939, resembles *O. levigata* in certain feactures. It has very narrow leaves like var. *scitula*, with white midribs and a lax inflorescence, the stems having much diffuse red and many large red papillae, stem tips bent horizontal, the style extremely short in flower, as in *O. levigata*. It agrees with *O. levigata* var. *rubripunctata* in having a bent stem tip, terminal sepal tips and very large fruits. The red papillae on ovary and sepals are light-sensitive, a character most strongly expressed in var. *rubripunctata* and in *O. ammophiloides* var. *laurensis*. The latter also has strongly bent stem tips, The strain from Les Escoumains thus combines characters from different sources and cannot be regarded as a variety of any one species, but rather as a hybrid of complex origin.

O. angustissima is clearly related to *O. argillicola* in being glabrate with very narrow leaves. The GATES & BARTLETT strains, both from Ithaca, N.Y., differ only in the longer leaves and slightly larger flowers and fruits of the former strain. *O. angustissima* can be regarded as a small-flowered descendant of the large-flowered *O. argillicola*. The plants were grown in the different climates of England and Maryland, which may acount for some of the size differences.

O. argillicola, endemic to the shale barrens of the Appalachian Mountains from Roanoke, Virginia, to Huntingdon, Pennsylvania, in West Virginia shows marked genetic similarity to *O. Hookeri* in California. It is isogamous but heterogametic, the complexes being *dilatans* ♀♂ *angustans* ♀♂ (RENNER 1933). MICKAN (1936) shows that these complexes are also related to *flectens* of *O. cruciata*, but they are clearly different although they differ in only one translocation, the chromosome ends being as follows:

angustans	1.4	2.3	5.6	7.8	9.10	11.12	13.14
dilatans	1.4	2.3	5.6	7.9	8.10	11.12	13.14

In the combination *angustans·flavens* from the cross *O. argillicola* X *O. suaveolens* 7 pairs of chromosomes are formed, indicating that all ends are the same in both. *O. argillicola* could thus be one of the parents of *O. suaveolens*. In the cross *O. argillicola* X *O. Hookeri* (small-flowered type), one type of seedling *(Hookeri-angustatae)* is viable. Similarly in *O. Lamarckiana* X *O. argillicola* the *laeta* type of seedling is lethal, the *velutina* type alone developing; and in *O. biennis* X *O. argillicola* the *rubrae* die as seedlings.

CLELAND (1937) described a homozygous strain of *O. argillicola* from Huntingdon, Pa. with 7 free pairs of chromosomes. RENNER (1941) found this indistinguishable from the West Virginia form and locates several genes of *O. argillicola* in particular chromosomes. CLELAND finds certain differences in the chromosome "ends" of the Pennsylvania form.

In more recent studies of *O. argillicola*, STINSON (1953) finds 29 races with 7_{11}, 27 races with ④, and a few strains having catenation up to ⑩. There is an almost complete absence of balanced lethals, and the chromosome ends are similar to those of *O. Hookeri* although the two species are phenotypically very different. The larger chromosome rings in some races of *O. argillicola* are evidently the result of crossing with other species.

Although *O. argillicola* & *O. Hookeri (O. franciscana)* are separated by the width of the continent, they are at practically the same beginning stage of catenation. One must assume that catenation in Oenothera began as a new phase of evolution, probably during the Pleistocene. Previously the differentiation of the ancestors of *O. Hookeri, O. grandiflora* and *O. argillicola* must have taken place by the same methods as in other genera of plants with free pairs of chromosomes. Since *O. argillicola*, which differs from *O. Hookeri* by at least three interchanges of chromosome ends, is homozygous, it is presumably derived from earlier ancestors with chromosome catenation in which many mutational changes in phenotype had already taken place. CLELAND (1937) regards *O. argillicola* as closely related to the Californian *O. Hookeri*, but this relation appears to have been exaggerated. The relation to *curvans* of *O. muricata* he regards as more distant. In crosses with *O. grandiflora (truncans· acuens)* he found in the combination *acuens · h argillicola* ④ indicating a close relationship between these two species, i.e. only one interchange of chromosome ends. On the other hand, *trucans· curvans* from the cross *O. grandiflora* X *O. muricata* showed ⑩²⁴.

MacDOUGAL (1905) pointed out that in *O. argillicola* the style elongates rapidly in the morning before the flower opens, thus exposing the spreading stigma lobes 3 or 4 mm above the unopened buds. The stigma could thus receive foreign pollen in the morning when the flower only opens that evening. Similar behaviour occurs in various other species, but only at the end of the season.

O. grandiflora

In *O. grandiflora* with ⑫ or ⑭ and in *O. Lamarckiana* with ⑫ the catenation is far advanced compared with *O. argillicola* and *O. Hookeri*, but the large flowers have been retained, and the "early" seasonal development of the south in *O. grandiflora*. STEINER (1952) finds that the races of *O. grandiflora* from Georgia are *biennis-grandiflora* in type, whilst those from Alabama belong to CLELAND'S *biennis I*. *O. grandiflora* shows the ancestral segmental arrangement of chromosome ends, so *O. grandiflora*, *O. Hookeri* and *O. argillicola* can now all be regarded as relict species from an earlier period of evolution. More recent seed collections from Virginia and North Carolina (STINSON & STEINER 1955) yield 8 strains, all but one having ⑭. Five of these are placed in the *biennis* group, the other three having a complex connected with *O. grandiflora*.

O. Victorini and *O. pycnocarpa* appear to be unrelated. The latter species occurs widely in New York State, and *O. Victorini* has a more northerly distribution, mostly in Ontario and Quebec. The measurements show no particular similarity. The specific differences are listed in GATES (1936, p. 320, Table XX). *O. pycnocarpa* var. *parviflora* from Hamilton and Georgetown, N.Y., differed in being smaller plants with smaller flowers and red papillae on the stem. *O. nutans*, described from Ithaca, N.Y., differs from *O. pycnocarpa* most conspicuously in having bracts which are caducous when the

---

²⁴ A specimen in the Gray Herbarium from Blue Ridge (Bedford Co.) Virginia, 1947, with petals 25 mm, ovary 10 mm, hypanthium 40 mm, bud cone yellow, glabrate, 15 mm, sepal tips 2 mm, midleaf 8 cm × 16 mm, has wider leaves than *O. argillicola* and may also be related to *O. grandiflora*.

flower wilts. The rosette leaves are crinkled, with liver spots and reddish midribs. The petals are delicate, the flower nodding and quickly wilting after anthesis.

Six strains of *O. deflexa* were grown in 1938 and 1939 from Riverside Drive near Windsor, Ontario. The differences between them are shown in Table VII.

It will be seen that *O. deflexa* has an extraordinary range of petal-length, from 9-20 mm, but in each culture the variation is narrow, so that the differences strike the eye at once.

Table VII. Characters of *O. deflexa* strains

Culture	Height	Leaves	Midrib	Midleaf		Petals	Buds
	in.			cm	mm	mm	
71.39	70	slightly crinkled	pink	17 ×	45	15	fine puberulence
72.	61	smooth	faint pink	18 ×	39	9	fine puberulence
73.	63	smooth	white	14 ×	38	20	subglabrous
74.	62	smooth	faint pink	17 ×	29	17	glabrous
75.	64	slightly crinkled	pale pink	17 ×	49	12–13	sl. hairy
76.	69	finely crinkled	faint pink	19 ×	44	17	sl. hairy
76A*	60	smooth	pale pink	20 ×	58	8–11	fine puberulence

* *O. deflexa* var. *bracteata*

Viewing these observations on variation in Oenothera cultures as a whole, they constitute the most detailed comparative measurements and observations yet made in the genus. They show beyond doubt that (1) each culture from the seeds of a self-pollinating plant breeds true within very narrow limits except for occasional trisomic or tetraploid mutations, (2) strains of a particular species in a limited locality show marked differences in such features as flower size, leaf width or crinkling, and pigmentation of particular organs. These appear superficially to represent Mendelian segregation, but owing to chromosome catenation this cannot be the real explanation. They rather represent a more recondite distribution of genic differences which has arisen earlier through mutation. (3) These variations generally do not trespass on the integrity of the species, but may occasionally show a cline of distribution running in a particular direction. (4) That certain species are of recent hybrid origin is shown by their geographic and phenotypic relations to two other species, but the existence of genotype complexes may prevent the recognition of such cases. The genus has specialized in chromosome catenation with self-pollination. Its multiformity is thus quite different from that of apomictic genera like Taraxacum or Hieracium, or of many other genera in which polyploidy or amphiploidy has been of frequent occurrence.

In certain respects the self-pollinating, true-breeding species of

Oenothera resemble the condition found in such apomictic genera as
Taraxacum or Parthenium. The strains are, however, probably much
more heterozygous in Oenothera, and yet both types of reproduc-
tion agree in producing a multitude of nearly related strains. CLAU-
SEN (1954) has discussed the evolutionary significance of partially
apomictic species. But facultative apomicts store their variability
in a different way from these Oenotheras, and being less heterozygous are probably less versatile in their response to environmental
changes. From the variability already described, one must assume
that the linked chromosomes in Oenothera undergo many exchanges
of several kinds during meiosis.

Naturalization of Species in Europe

Before taking up further problems of relationship and phylogeny
within the genus, we may consider the forms which have been
spreading over Europe in post-Columbian times. After a careful and
detailed historical study of herbarium specimens and other botanical
records, WEIN (1936) concludes that *O. muricata, O. biennis, O.
angustissima* and *O. parviflora* were all introduced during the 17th
century. The first Oenothera figured, by ALPINO (1627), he identi-
fies as *O. muricata*[25], which came from England to Padua in 1612 and
was probably derived from the American coast after 1606. The
plant of PARKINSON's "Theatrum Botanicum" is identified as belonging to *O. biennis* in the general sense. *O. muricata* with white
midribs is found to have been grown in England before 1612, reaching Italy in 1612, Switzerland 1619, Germany 1627, Holland 1633,
Denmark 1642, Sweden 1658, chiefly by transmission from one Botanic Garden to another. *O. biennis* was probably the first Oenothera
to be grown at the Paris Botanic Garden, 1623. There it was assumed to be the same as the form described by C. BAUHIN and P.
ALPINUS, which WEIN identified as *O. muricata*. Evidence is obtained
that *O. biennis* reached Holland in 1633, England in 1648, Germany
1660, Italy 1662, Sweden 1685, Alsace 1691, Switzerland in 1715 and
Russia in 1736. It thus reached its widest distribution in European
gardens during the second half of the 17th and the first half of the
18th century, displacing *O. muricata*. *O. muricata* with red midribs
was first recorded in France, 1688, entered Italy in 1723, England

---

[25] In a later study, RENNER (1942b) finds 12 species and forms now naturalized in
Germany. He gives good reasons for identifying ALPINO's plant as *O. parviflora*, and
has determined the complexes of eleven species. He also (RENNER 1943b) gives
localities in Germany for *O. ammophila, O. parviflora* and *O. rubricaulis*, and describes
O. silesiaca from Boberufer bei Nauburg and other localities. BAERECKE (1944) has
worked out chromosome formulae for the above four species as well as for *O. Bauri*
and a species, *O. Beckeri* RENNER, found on the Holland border. RENNER (1945) has
made a fuller analysis of the *percurvans* pollen complex of *O. ammophila* FOCKE, and
gives (RENNER 1951) a further discussion of some European species.

1739, Germany 1753. WEIN thinks it was introduced afresh from
North America as a new race, but it remained everywhere uncommon.
O. angustissima is recognized as grown in France in 1659, England
1683, Holland 1687, Italy 1689, Germany 1700. It was commoner
in gardens than *O. muricata*. WEIN finds evidence that *O. angustissima* came to France from her North American Colonies. In that case
it would have been a form belonging to var. *quebecensis* introduced
from the St. Lawrence.

O. biennis sulfurea appeared as a mutation in the Botanic Garden
at Leyden between 1668 and 1686. *O. biennis* ceased to be a garden
plant in the 17th century. The first evidence of its escape and naturalization is in Holland 1683, then in Germany 1711, Italy 1726,
Alsace 1728, Poland 1730, Switzerland 1742, and France 1748.
It was first naturalized along rivers, including the Elbe (1711),
Saale, Oder, Rhine, and Etsch (1754). The first certain record
showing naturalization of *O. muricata* was on the Danube in 1728[26].

WEIN does not attempt to trace the history of *O. grandiflora*,
O. Lamarckiana or *O. suaveolens*, but they must be considered
briefly here. The nearly glabrous *O. grandiflora* was discovered by
William BARTRAM near Tensaw, Alabama. He sent seeds to Dr.
JOHN FOTHERGILL in London in 1776, and the species was described
by SOLANDER in Aiton's *Hortus Kewensis* in 1789, from plants grown
at Kew, as related by DAVIS (1926). Although it does not escape
from gardens in England and America as *O. Lamarckiana* has done,
yet it flourishes on the Lancashire coast along with *O. Lamarckiana*
and *O. biennis* (GATES 1914a). *O. suaveolens* DESF., *albicans* ♀·
flavens ♂ (♀), has a ring of 12 chromosomes. The *albicans* complex
agrees with that of *O. biennis* (RENNER 1925) and *flavens* is very
similar to *acuens* of *O. grandiflora*. There can be little doubt that
O. suaveolens has arisen in France, perhaps as a hybrid between these
two species; but the hybrid catenations indicate an even closer
relationship to *O. argillicola* (see p. 86). DE VRIES (1918) and others
found *O. suaveolens* in quantity near Fontainebleau unmixed with
O.biennis or hybrids. He found half the seeds empty, as in *O. Lamarckiana*, owing to balanced lethals. The cross which gave rise to it is
most likely to have occurred in a botanic garden where different
forms were grown together.

Much has been written regarding the origin and history of *O.
Lamarckiana*. DAVIS (1916, 1921, 1926) has long contended that
it is of hybrid origin in Europe. From crosses between *O. franciscana*
and *O. biennis* he produced in F_3 a segregate, *O. neo-Lamarckiana*,
which resembled the plant of DE VRIES' experiments in many respects. It was selfed for seven generations, but it throws a *franciscana*

---

[26] GECKLER (1950) has cultivated a form practically identical with *O. muricata*
from near Sunbury, Pennsylvania. Its α chromosome complex is the same as the
rigens complex of *O. muricata*.

type. It also produces a dwarf type, an *albida*-like mutation, and triploids. The seeds are 90% fertile, those of *Lamarckiana* 35% fertile. *O. neo-Lamarckiana* is described by HOEPPENER & RENNER (1929) as *O. (franciscana* X *biennis) rubefacta* with large flowers and white midribs. In crosses, *O. franciscana* behaves exactly like *O. Hookeri*. As CATCHESIDE (1940) shows, the *gaudens* complex of *Lamarckiana* and the *rubens* of *biennis* have the same structure of chromosome ends (see below), so that *Lamarckiana* might have arisen as a hybrid of *O. biennis* with some form of *O. Hookeri* or another large-flowered species. It is now widely distributed as a garden escape in England, but that it originated there as late as 1860, when Messrs. CARTER announced their introduction "from Texas" is very difficult to believe. DAVIS (1916) suggested that it arose on the Lancashire sand hills by a cross between *O. biennis* and *O. franciscana*. In a study of these plants (GATES 1914a), colonies of *O. biennis* and *O. Lamarckiana* were found, as well as hybrids and plants with large flowers and slender *O. grandiflora* buds and other characters indicating that at some time all three species had reached this area. Oenotheras are known to have been flourishing there as early as 1805, but whence or when they arrived is unknown.

VAIL (in MACDOUGAL, VAIL & SHULL 1907) cites a specimen in the Gray Herbarium, collected wild on the pine barrens of New Jersey, having large flowers but the stout buds of *O. Lamarckiana* rather than the slender ones of *O. grandiflora*. A recent personal examination of this interesting specimen shows that it is undoubtedly *O. Lamarckiana*. The sheet bears the words, "Pine barrens of New Jersey. "Wild" Miss Treat of Vineland 1871." and below is the annotation in pencil, "Oen. original of Lamarckiana??? B.L.R." (obinson). The only question is whether it really grew wild as stated. The petals are 45-50 mm, sepals ± hairy, the stem bears many red papillae and a midleaf measures 11 cm x 23 mm. The buds are greenish and the midribs were red, as in mut. *rubrinervis*. A sheet marked "Hort. Cantab. 1862, e semina Thompson, Norwich" is also *Lamarckiana*. In pencil is added "said by Engl. horticul. to come from Texas." It shows the interest which was being taken in the species at that time.

To show the propensity of *O. Lamarckiana* to escape from cultivation, the following specimens in the Gray Herbarium are cited: Kabul, Afghanistan, 1939. Lotbinière, Quebec, 1932 (discovered by Prof. MARIE-VICTORIN and myself growing by the roadside close to a potato field). Annapolis Royal, N.S., 1921. Carleton, Yarmouth Co., 1921. Yarmouth, N.S., 1920. Orono, Maine, 1892. Cambridge, Mass., 1880. Ancho Bay, Mendocino Co., California, 1938. Arcata, Humboldt Co., Calif., 1936. Fort Lewis, Washington., 1937. Pangborne Lake, Washington, 1939. Victoria, B.C., 1918. The Yarmouth, 1920, specimen was probably a hybrid with the local *O. novae-scotiae*. It had petals 25 mm and a short style. A specimen from North Worcester, Mass., 1920 ("in a cultivated field"), may be a hybrid of *O. Lamarckiana*. It has petals 35 mm and very narrow leaves (midleaf 9 cm x 10 mm). Another, from Hudson Falls, Wash. Co., N.Y. has petals 35 mm, hypanthia very long (50 mm), ovary 10 mm, midleaf 11 cm x 23 mm. A specimen from an old cellar in a deserted yard at Independence, Oregon, 1918, probably belongs to *O. Hookeri*. It has petals 30-35 mm, red papillae only at apex of stem; ovary hairy, with red papillae; bud cone reddish, with no red papillae, sepal tips only 1-2 mm; midleaf 6.5 cm x 10 mm. At Barss Corner, (Lunenburg Co.) N.S., *O. Lamarckiana* was found to be spreading even in greensward (GATES 1936, p. 347). A specimen in the Jardin des Plantes, collected by MICHAUX in North America (without further locality) in the 18th century, differs from the *Lamarckiana* of DE VRIES mainly in having stem-leaf petioles. The mystery as to when and where *O. Lamarckiana* originated appears to be still unsolved. However, it is no longer of great consequence, as the origin of new hybrid species is now recognized to occur both in cultivation and in the wild.

O. purpurata was described as a new species by KLEBAHN (1925) in a study of the Oenotheras on the Lunenburger Heide of north-western Germany. Seeds collected at BEVENSEN in 1912 from a single self-pollinating plant which appeared to be *O. biennis* X *biennis cruciata* were grown. In the descendants in 1914 appeared a single plant which was afterwards called *purpurata*. The stem was strongly red striped to violet, especially in the upper part, with red papillae and soft pubescence, the stem tip bent. It was late flowering, the midribs mostly whitish; ovary green, with small red papillae; hypanthium 30-35., sepals with two broad red stripes, petals ca. 28 x 28 mm, stigma often above the anthers in the bud.

This strain bred true and was found (RUDLOFF, 1929) to have seven free pairs of chromosomes. Seeds sent to the writer produced a uniform race which, in London, was strongly biennial, the stems produced in the second year being stout and much fasciated. It showed considerable resemblance to *O. Hookeri* but with smaller flowers. There was no red pigmentation until flowering, when a red flush developed from the stem base upwards. The lack of catenation was confirmed (GATES & GOODWIN 1931) and the cytology of *purpurata* was compared with that of *blandina*, a homozygous secondary mutation from *O. Lamarckiana*, also with seven free pairs. Both are homozygous derivatives from ancestors with catenation. It was found that they differed mainly in the amount of interlocking of the ring pairs, *purpurata* having all the rings free in 57% and *blandina* in 26% of the pollen mother cells. The view was also expressed that the terminal portions only of homologous chromosomes in Oenothera pair side-by-side, the longer median portions remaining unpaired (asynaptic). This will be referred to again later.

GENETICS AND CYTOLOGY

Flower-size inheritance

From crosses between *O. purpurata* and *O. Hookeri*, RUDLOFF (1929) found that the reciprocals differed only in flower size. In *purpurata* the petals were *ca.* 22 x 25 mm, in *Hookeri ca.* 45 x 36 mm. The F_1 of *purpurata* x *Hookeri*, with 90% germination, gave 38 plants, 8 of which had *Hookeri* flowers (petals 43 x 34 - 46 x 36 mm). The rest were much smaller, ranging from 15 x 15 - 24 x 21 mm (75 measurements), the smallest flowers being much smaller than in *purpurata*. In F_2 there was no clear segregation and the pure parental types did not reappear. Numerous other petal measurements in the reciprocal and in back-crosses led to the conclusion that there was clear monohybrid splitting for flower size, but no marked segregation in other characters, the differences in foliage being polymeric. *Purpurata* has narrower bracts than *Hookeri*. Whether the smallest flowers were confined to certain plants is not clear. Reciprocal F_1 hybrids showed a difference in flower-size only, *purpurata* X *Hookeri* having petals 24 x 26 mm, *Hookeri* X *purpurata* 20 x 22 mm. HOEPPENER & RENNER (1929) found that in crosses between two other homozygous forms, *O. Hookeri* and *deserens* (a secondary mutation from *O. Lamarckiana*), the reciprocal hybrids again differ only in flower-size. In both these cases the difference must be referable to the plasma. KÖHLER (1929) obtained similar results in reciprocal crosses between several species of Epilobium, and in maize RHOADES (1933) shows that the egg cytoplasm plays the chief role in the expression of male sterility. Much other evidence goes to show that petal length is determined by plasmatic as well as nuclear differences, and that flower-size is in a class by itself among the inherited characters of plants (see p. 39). MICHAELIS (1949, 1951, 1953) has made extensive studies of plasmatic effects and plasmon changes in hybrids of Epilobium species.

Inheritance of petal-size was studied on a large scale in four generations of reciprocal hybrids between *O. biennis* and *O. rubricalyx* (a dominant gene mutation from *O. rubrinervis*) by GATES (1923). The results showed a type of variability in some respects unique, with large and small flowers on the same plant, and even long and short petals in the same flower, as well as petals which developed slits, apparently from crumpling in the bud and the resulting interference of petals with each other's growth. It is necessary to conclude that flower-size is partly plasmatic in its determination, at least in Oenothera, yet the flower-size of any strain remains relatively constant and large flowers, for example, are still formed even under conditions of starvation. EMERSON & STURTEVANT (1932) showed

94

that the *brevistylis* factor *br*, like the *Co* factors for flower size, is independent of all the genes, even when there is a ring of 14 chromosomes, but they found *Co* and *br* "linked", with 15% of crossing-over: They concluded that these were genes in the same chromosome far from the translocation point, but that view is no longer tenable since it was found that reciprocal hybrids between homozygous forms differ in flower-size. Some Oenothera hybrids develope pale green patches on the leaves, due to incompatibility of mitochondria, chloroplasts and cytoplasm in certain species hybrids, the plastids in some species evidently having specific characters (RENNER 1924, 1937b).

As there is no record of *O. Hookeri* naturalized in Germany (it would probably not withstand the winters) it is probable that *purpurata* got its *Hookeri*-like characters as a parallel development. The derivation of this homozygous form from *biennis*-like ancestors by a mutation implies a very considerable rearrangement of chromosome ends.

In an account of the genetics of the wild Oenotheras of Northern Germany, RENNER (1937) finds *O. biennis* everywhere uniform from Upper Bavaria to the East Baltic Coast. *O. syrticola* BARTL. (known as *O. muricata*, petals 13 x 13 mm) is on the Elbe; and on the North Friesian Islands, *O. ammophila* FOCKE. *O. germanica* BOEDIJN (1924) from near Berlin is treated as a subvariety *rhodoneura* of *O. ammophila*. The egg complexes of *ammophila* and *syrticola* are practically identical *(rigens)* but the pollen complexes differ. *Curvans* of *syrticola* includes gene M *(marginata)*, i.e. red margin of leaves and bracts and slight red in the young inflorescence. *Percurvans* of *O. ammophila* has genes for red stem papillae, very strong bending of the stem tips, and strong growth. *Rigens* also has a gene for red papillae (P) but it is not identical with that of *percurvans*. *O. ammophila* differs in having one free pair of chromosomes (SHEFFIELD 1927), but a ring of 14 was found in one pollen mother cell.

O. parviflora L. is in Schleswig and probably in Freiburg-in-Br. and in Berlin. According to RUDLOFF, *O. pachycarpa* RENNER is a synonym. *O. rubricaulis* KLEBAHN (1925) is heterogamous, having the complexes *tingens* ♀· *rubens* ♂ and catenation ⑥ + ⑧. This is the same as in *O. biennis*, which is evidently one of the parents.

O. Bauri BOEDIJN (1924) was described from Friedrichshagen bei Berlin. The identical form has long been on the upper reaches of the Weichsel. It is also heterogamous, its complexes *laxans·undans*. Near Dantzig it crosses with *O. rubricaulis* ♀, producing a hybrid *(O. Weinii, tingens·undans)* which grows as an apparent variety of *O. Bauri*. Varieties which differ in midrib-colour or flower-size are often products of crossing. *O. Lamarckiana* occurs in places under the name *O. grandiflora*. South of Berlin are mixed populations of *O. biennis, ammophila* and var. *germanica* with their hybrids. *O. (biennis* X *ammophila) albipercurva* is constant and thrives in large stands as a "new species". A variety of this, lacking stem papillae, is equally successful at Luckenwalde and may have originated as a mutation. *O. biennis* var. *parviflora* = *O. rubricaulis* KLEBAHN (1925) = *O. muricata* var. *latifolia* ASCHERSON.

O. ammophila has long been known from the sand dunes on the southeast coast of the North Sea. It is very much like *O. muricata*. Both species have the same egg complex, *rigens*, whilst the *percurvans* of *O. ammophila* differs from *curvans* of *O. muricata* in producing more extreme (crozier-like) bending of the stem tip, *curvans* determining also the red papillae on stem, ovary and sepals of *O. muricata*.

It is thus clear that since being transported to Europe by various means the Oenotheras have continued to spread, crossing occasionally and thus producing new and constant forms, as well as producing occasional mutations such as the gene, R, for red midribs in *O. muricata*, *O. ammophila* and *O. biennis*. RENNER (1921) found this gene in the complexes *rubens*, *gaudens*, *velans*, *rigens* and *hookeri*. HERIBERT-NILSSON (1915) originally showed that R was a zygolethal in *O. Lamarckiana*. Homozygotes (RR) were eliminated, and in crosses between two Rr individuals with red midribs the ratio red: white was 2 : 1, not 3 : 1.

Rapid evolutionary change is thus occurring in Europe as in America. The view was early expressed (GATES 1911, p. 605) that crossing could lead to germinal instability. "It seems that the mutation phenomena in *O. Lamarckiana* are due to a disturbed or unstable condition of the germ plasm, which has probably resulted from crossing in the ancestry." With all the modern refinements of genetics that view still holds, and it is reasonable to suppose that even gene mutations may sometimes accompany or follow a cross which has introduced new combinations into the genome. That mutation is not merely hybrid splitting was, however, clearly shown (GATES 1914b) by comparing the origin of mut. *rubricalyx* with reciprocal hybrids of *O. grandiflora* X *rubricalyx*. Among 84 plants derived from the cross *O. syrticola* X *rubeflexa*, RENNER (1937) obtained one plant with diffuse red calyx. This dominant mutation apparently resembles *O. rubricalyx*.

The extensive work of CLELAND (1950) has made progress towards the determination of lines of phylogeny in Oenothera by the study of the pairing of chromosome ends. He suggests (p. 244) that "the term 'species' can be properly applied to any group of individuals of common origin and occupying a definite geographical range, which possess a high degree of structural similarity, and which at the same time displays a recognizable distinctness from other similar groups." With such a definition, it seems clear that harmonization of the views derived from the study of chromosomal interchange and those of the advanced taxonomist will ultimately be achieved.

Development of Oenothera Genetics

The half-century and more of work with Oenothera, beginning with the epoch-making investigations of DE VRIES, has seen the unfolding of one process after another relating to the genetics and evolution of the genus. DE VRIES' conception of new species arising at one stroke was clearly substantiated only in the case of the mutant *gigas*, which served as the starting point for the whole field of polyploidy (GATES 1909a). Mutation *gigas* was shown to be a cell giant, with larger nuclei and larger cells whose shape was frequently altered, the increase in one dimension being greater than in another. Many species in other genera must have originated as tetraploid mutations, but in Oenothera tetraploid mutations have never succeeded in nature, although they must occur from time to time, as several species have produced them in culture. Natural polyploidy is found, however, in the genera Kneiffia and Hartmannia (HECHT 1942, HAGEN 1950), which are nearly related to Oenothera.

Trisomic mutations were the next to be recognized and proved (GATES & THOMAS 1914). These are also recorded in many species of Oenothera and other genera but, having an unbalanced chromosome number, they are generally pollen-sterile and can only be expected to survive under exceptional conditions such as apomicty. The first case of parallel mutations was discovered in *O. biennis* mut. *lata* (GATES 1912a) and this important principle has since been widely applied in plants and animals (GATES 1920). VAVILOV later (1922) applied the term "the law of homologous series" to such phenomena.

The publications of RENNER (1925 etc.) showed that all Oenothera species except the few which are homozygous are composed of two complexes, the heterozygous combination of these two being viable while the diploid homozygote of each complex is lethal, as shown by the presence of 50% of empty seeds (zygolethal) and frequently of 50% of non-viable pollen grains (gamete lethal). RENNER (1925), however, is skeptical of gamete lethals. In many species-crosses part or all of the seedlings may die in the cotyledon stage. In a zygote lethal death occurs still earlier, in seed (embryo) formation. The balanced lethal condition has evidently arisen through the transformation of a seedling-lethal into a seed (embryo) lethal by a mutation. RENNER (1933) has shown that the various lethal combinations stop development at different stages; e.g. *flavens·flavens* in *O. suaveolens* developes to ca. 200 cells, *velans·-velans* in *O. Lamarckiana* is still smaller, whilst the *gaudens·gaudens* embryo is often only two cells. RENNER shows many other degrees of lethal effect; e.g. the pollen complex *flectens* of *O. cruciata* NUTT. contains the genes *Fl* (bent stem), *M* (margined leaves), *lor* (very narrow leaves) and *Pil* (fine-haired calyx). Plants which are *lor·lor* are cripples with short, thin stems and thong-like leaves. *Fl·Fl, Lor·lor* plants are still weaker, small pale rosettes seldom forming a stem. *Fl·Fl, lor lor* dies as a seedling, while *Pil·Pil* kills still earlier, in the embryo. That a graded series can also occur in a single hybrid type, was shown in the case of *O. rubricalyx* X *eriensis* (p. 42). DAVIS (1947) found a case of mutational change from cotyledon lethal to seed lethal (p. 13), which indicates how the condition of balanced lethals may have arisen. A less extreme condition than that of balanced lethals has been found in various species of *Drosophila* (DOBZHANSKY 1949) in which the heterozygote of certain chromosome inversions is more viable under certain conditions than either homozygote.

The cytological investigations developed in the meantime. The earlier observations of GATES showed many of the 14 chromosomes connected end-to-end. Then CLELAND (1922) found in *O. franciscana* a ring of only four chromosomes, the rest being in free pairs. Various other fixed linkages of the chromosomes, including a ring of 14, were soon discovered in other species and hybrids, the term catenation being applied to this arrangement (GATES 1931). BELLING & BLAKESLEE (1926) put forward the theory of segmental interchange to account for chromosome linkage in Datura. This was soon applied to Oenothera (CLELAND & BLAKESLEE 1930, 1931) and successful predictions were made regarding the catenation in various hybrids, the system of numbering the ends 1-14 being introduced. It was pointed out that in Datura the chromosome rings are small and temporary (in cultivated hybrids) while in Oenothera they are permanently maintained by the presence of balanced lethals. A ring of 4 can arise from a single interchange of chromosome arms, as it has done in Pisum and other genera. In Oenothera the ring probably grew in size through crossing as well, a cotyledon-lethal hybrid being converted into a balanced seed-lethal at the same time.

CATCHESIDE (1933) concluded that chiasma formation was concentrated towards the ends of chromosomes, even in the free bivalents, and EMERSON (1936b) showed that trisomic mutations were to be expected following certain irregular configurations in the zigzag arrangement of chromosomes at the first meiotic metaphase. CATCHESIDE (1936) showed that in *O. Lamarckiana* with ⑫ a total of 73 trisomics was theoretically possible, there being three possible types of non-disjunction to yield 8-chromosome gametes. FORD(1936) showed that in forms with ⑭ non-disjunction can occur in four cytologically distinguishable ways, producing a total of 98 possible 15-chromosome zygotes.

End-to-end catenation of the chromosomes naturally supported the then current view of telosynapsis. DARLINGTON (1931), in a detailed study, brought forward evidence of parasynapsis, which was substantiated by CATCHESIDE (1931). In further studies, CATCHESIDE (1933) found that while chromosome pairs are potentially capable of pairing throughout their length, yet in rings only the end segments actually pair, the differential middle segments (which are believed to distinguish mainly the two complexes of a species) remaining normally unpaired. There is evidence that these end segments are of different lengths in different pairs of chromosomes. The chromosomes also (MARQUARDT 1938, BHADURI 1940) form a graded series in length, ranging from *ca.* 18-10 units. The numerous genes now recognized by RENNER (1925), EMERSON & STURTEVANT (1932) and others may be briefly listed here.

> Mm = red or green margined leaf.
> Pp = red papillae on sepals. $\left.\begin{matrix} \\ \end{matrix}\right\}$ interrelated
> Str, str = stripes on sepals.
> Nn = tall or dwarf.
> Ss = yellow or sulphur petals.
> Rr = red or white midribs (RR lethal).
> $B_1B_2b_1b_2b_3$ = leaf width.
> Sp sp = pointed or blunt leaves, sepals, buds.
> Cu cu = bent or erect stem tip (Cu Cu lethal).
> Co, Co_1, Co_2, co_1, co_2 = decreasing flower size.
> Spr spr = spreading or non-spreading sepal tips.

Many of these have been placed in their complexes or even in particular chromosome arms (See RENNER 1942). They determine characters such as red or white midribs, erect or bent stem, red or white papillae on the stem or other parts, broad or narrow leaves, petal colour and width, subterminal sepal tips, tall of dwarf stem and other features.

Most of these genes are in the pairing end-segments of chromosomes, since they are occasionally transferred from one complex to another (by crossing-over) without altering the end-homologies. EMERSON (1936) points out that the species of Oenothera have nevertheless not become completely homozygous for these genes, probably because the species are so recent.

Oenothera Cytology

There is much evidence that segmental interchange has taken place mostly at or near the centromere, and there seems no reason for regarding the median unpaired segment of the chromosomes in a ring as (in some cases) even in any close sense homologous. To explain the pairing of ends it is necessary to assume 14 sets of specific attractions, whether the chromosomes are in pairs or in a ring of fourteen. The evidence indicates that when the middle portions pair it is only a zipper effect, the specific attractions being confined to rather short end-segments. The frequent interlinking of free ring-pairs appears to result from the fact that different specific end-pairs are "seeking each other out" at the same time and thus become entangled. For these specific attractions there is at present no physical explanation. Interlinking of pairs does not occur in non-catenating organisms, and it is only possible when pairing begins at, or is confined to, the free chromosome ends.

Since, as already stated, the chromosome pairs do form a graded length series, segmental interchanges, even if they all occur at the centromere, will produce chromosomes with slightly unequal arms.

As pointed out many years ago (GATES & NANDI 1935) in a discussion of this subject, the method of chromosome pairing in Oenothera is thus strictly neither telosynapsis nor parasynapsis but acrosyndesis in the literal sense. The tips of the chromosomes pair side-by-side while the median portions, which constitute the bulk of the chromosome, show asynapsis and actual repulsion during most of the critical stages of pairing. All observers are agreed that pairing in Oenothera begins at the ends of the chromosomes, and a critical reading of the evidence leads to the conclusion that it is frequently confined to the ends, only extending further as a mechanical zipper effect. In some plants the centromeres appear to be the original centres of the attraction which leads to synaptic pairing. In Oenothera this is certainly not the case, but they are rather centres of repulsion between homologous portions of chromosomes. The studies of GATES & GOODMAN (1931) on *O. purpurata* and *O. blandina*, of HEDYATULLAH (1933) on *O. missouriensis* and of WISNIEWSKA (1935) and MARQUARDT (1938) on *O. Hookeri* agree in showing that even in these homozygous forms chromosomes in the ring-pairs are attached only by their ends, strongly repelling in their middle portions. Indeed, this repulsion, only the chromosome ends remaining attached, is the most characteristic difference between syndesis in Onagraceae and in genera such as lilium in which members of a bivalent remain attached and twisted about each other throughout their length. It has been assumed, because of this lack of median pairing, that in catenated chromosomes the middle segments are differential, representing in the aggregate the difference between the two complexes or genomes, and that this is the reason for their failure to pair or to remain paired. But since the same repulsion occurs in homozygous forms with only free chromosome pairs it must have some purely physical non-specific cause.

WISNIEWSKA (1935), MARQUARDT (1938), and BHADURI (1940) agree in finding in resting somatic nuclei 14 chromocentres or pro-chromosomes, corresponding to the mid-portion of the chromosome around the centromere. WISNIEWSKA finds this mid-portion more condensed than the ends, the latter having (as CATCHESIDE 1931,

SIKKA 1940, PATHAK 1940, and others have observed) a terminal granule. Some of these terminal granules may represent unterminalized chiasmata, but terminalization is not a universal phenomenon. HAGA (1944) concludes that in Paris and Trillium the chiasmata do not move after they are formed. MARQUARDT distinguishes the heterochromatic (deeply staining) mid-segment from the euchromatic ends which remain paired. One may suppose that the presence of excess nucleic acid nullifies an attraction which might otherwise develop between the middle segments. WISNIEWSKA suggests that an exchange between non-homologous segments, leading to ring formation, depends on whether the chromosomes separate easily and whether the parts arising by interchange join easily with each other. It appears that only species in which the meiotic chromosome pairs form rings are capable of undergoing catenation, for chromosomes which remain paired and intertwined throughout their length obviously cannot separate easily or form rings. Among 100 pollen mother cells of *O. Hookeri* examined for linkage, WISNIEWSKA (1935) found one nucleus with a ring of 4 chromosomes.

All the evidence goes to show that in the homozygous Oenotheras the pairing, even when it takes place throughout the length of the chromosomes, is vary loose, and most definitely expressed at the ends. MARQUARDT found the mid-diplotene chiasma frequency in the euchromatin ends to be 2-4 per bivalent, and he was unable to find any certain chiasmata in the heterochromatin. Such terminalization of chiasmata as occurs must therefore arise in and be confined to the pairing ends. Incidentally, he found no suppression of the chromosome split in the last pre-meiotic division, contrary to the precocity theory of meiosis. In the hybrid *O. flavens. h. Hookeri*, MARQUARDT found a ring of 4, and from differences in the lengths of the chromosomes in the ring he concluded that besides a reciprocal translocation there was a deletion of heterochromatin near the centromere. He also concluded that the pairing of "homologous" heterochromatic parts (chromosome middles) took place in early leptotene, the euchromatic ends only pairing in zygotene, but this is apparently erroneous. The evidence indicates on the contrary, as we have seen, that pairing begins at the ends and only proceeds to the middle as a necessary effect of both ends being paired, the pachytene pairing being very loose.

MARQUARDT, in his study of *O. Hookeri*, was only able to detect one pair of chromosomes with satellites, producing nucleoli, but BHADURI (1940, 1948), in a critical study of the somatic chromosomes of various Oenothera species and hybrids, showed that they all have at least two pairs of sat-chromosomes, producing four nucleoli, In *O. Hookeri* one pair of nucleoli is large, the other very small, whereas in *O. Lamarckiana* and other species with catenation there are one large, one small and two intermediate nucleoli. When these fuse in somatic prophase, the four chromosomes by which they were formed remain attached to the single nucleolus formed by the fusion.

Frequently chains of chromosomes occur instead of rings. These can be explained as follows: In a form with ⑭* there will be two pairs of chromosomes, each attached to the nucleolus which they have produced at their satellite ends. These ends are attached, generally facing each other to the nucleolus they have produced, but not to each other. When the nucleolus dissolves, these two ends, with their satellites, will be left as free ends. In this way four free ends can arise instead of a ring of 14. Thus occasional free ends are not necessarily a sign of failure of chiasma formation. There is at present no evidence that the two pairs of chromosome ends which bear satellites ever form chiasmata.

The fact that in meiotic prophase all four nucleoli are generally fused into one, which thus has four chromosomes attached to it, may interfere with the later specific pairing of chromosome ends. Since species with seven free pairs of chromosomes also have two pairs of nucleoli, this latter condition is older in the genus Oenothera than catenation, but it does not necessarily prove that seven is a derived number, although pairing of segments in two chromosomes of a haploid Oenothera suggests that possibility. Many other features of nucleoli in relation to chromosomes have been discussed elsewhere (GATES 1942).

Certain other conclusions regarding the chromosomes of Oenothera are drawn from experimental work. In a haploid Oenothera, CATCHESIDE (1932) found pairing between certain of the seven chromosomes, which indicated reduplication of a terminal segment in one chromosome and of an interstitial segment in another. In a study of crossing-over within large chromosome rings, CLELAND & BRITTINGHAM (1934) found that three types of double cross-over can occur in a ring proximately to a given gene, on the assumption that every chiasma represents an antecedent cross-over. The gene for mut. *stenophylla*, which always arises from a trisomic such as *liquida* and not from *O. Lamarckiana* directly, is believed to be near the centromere. It is believed that genes close to the distal end of the pairing segments are probably transferred to the opposite complex by crossing-over often enough to render them apparently independent of their complexes.

By the action of X-rays on the pollen of the homozygous *O. blandina*, CATCHESIDE (1935) obtained a number of mutations in F_2, mostly with narrow leaves, in addition to five plants with ④, one

---

* This symbol means a ring of chromosomes. These symbols are used for all cases of catenation (e.g. ⑫, ⑭ etc.).

with ⑥ and one with ④ + ④. There was evidence of certain deletions. Chromosome interchanges in one plant produced a ring of 4 in which two chromosomes of normal length alternated with a very long and a very short one. The two latter had arms of unequal length, the short arms always being oriented in the same direction; thus confirming the hypothesis of fixed orientation of each chromosome in the ring, as well as fixed place in the ring. By crosses between the three derived homozygous forms, *O. deserens*, *O. blandina* and *O. purpurata*, forms were obtained (GATES & CATCHESIDE 1931) with a ring of four or a ring of six, thus showing that catenation can arise by crossing between forms with seven free pairs if they carry different segmental interchanges. Both segmental interchange and hybridization are thus doubtless involved in the rapid development of forms in which all the chromosomes have undergone segmental interchange.

Other cytological changes of relatively frequent occurrence will be crossing-over in the end segments and (less frequently) translocations between non-homologous chromosomes. RENNER (1943a) gives reasons for believing that translocations can arise from various forms of interlocking.

Specificity of chromosome ends

Coming now to the specific attractions of chromosome ends in Oenothera, CATCHESIDE (1940) has shown that the American system of numbering the ends and that of RENNER differ only slightly. CATCHESIDE's data bridge the gap and thus form the basis for future advance. CLELAND (1949, 1950) has since added other determinations which throw further light on the species relationships. Some of these end-numbers are given below, *O. Hookeri* being taken as a starting point.

ʰ *Hookeri*	1.2	3.4	5.6	7.8	9.10	11.12	13.14
ʰ *argillicola*	1.4	3.6	5.2	7.10	9.8	11.12	13.14

The hybrid between these two forms should then have ⑥ + ④ + 2_{11} These two widely separated species have only two chromosome pairs with similar ends, which is not surprising as they must have been evolving independently over probably at least the whole postglacial period. On the other hand, the complexes of *O. grandiflora* from Alabama, although they produce ⑭, are much more closely related to *O. argillicola*.

acuens	1.4	3.2	5.6	7.10	9.8	11.12	13.14
truncans	1.13	3.7	5.2	4.6	9.14	11.10	8.12

Acuens · argillicola should thus have ④ + 5_{11} and *truncans argillicola* ⑫ + 1_{11}. But *Hookeri·acuens* should have ④ + ④ + 3_{11} and *Hookeri·truncans* ⑭, showing the more distant relationship between *O. Hookeri* and *O. grandiflora*. These four catenations have all in fact been observed by CLELAND.

The complexes of *O. Lamarckiana* are

| velans | 1.2 | 3.4 | 5.8 | 6.7 | 9.10 | 11.12 | 13.14 |
| gaudens | 1.2 | 4.9 | 5.6 | 7.12 | 10.13 | 3.11 | 8.14 |

and CATCHESIDE shows that in mut. *lata* the extra chromosome is 5.6, while in *mut. pulla* it is 3.11. The hybrid *Hookeri· velans* should have $④ + 5_{11}$, showing the near relationship of *O. Hookeri* and *O. Lamarckiana*, while *Hookeri·gaudens* should have $⑩ + 2_{11}$. Both these catenations have in fact been observed by various investigators. The homozygous *O. purpurata* shows its relationship to *O. Hookeri*, since its ends are as below, like *h. franciscana* and *excellens*, whereas *blandina* ends are as shown.

| purpurata | 1.2 | 3.4 | 5.6 | 7.10 | 8.9 | 11.12 | 13.14 |
| blandina | 1.2 | 3.4 | 5.6 | 7.10 | 8.13 | 11.12 | 9.14 |

The hybrid *purpurata·blandina* should then have $④ + 5_{11}$, which GATES & CATCHESIDE (1931) observed.

As an example of a mutation, the ends are in *rubricalyx*

1.2 3.4 5.6 7.14 8.13 9.10 11.12

thus having four chromosomes in common with *velans* and two in common with *gaudens*. The *rubens* complex of *O. biennis* has the same structure as the *gaudens* of *O. Lamarckiana* (CATCHESIDE 1940). The locus of the *brevistylis* gene is found (CATCHESIDE 1954) to be in chromosome 11.12 near the distal end of arm 12.

Similarly, *O. suaveolens*[27] (naturalized in France, where it perhaps originated) is

| albicans | 1.4 | 2.14 | 3.6 | 5.7 | 8.9 | 10.12 | 11.13 |
| flavens | 1.4 | 2.3 | 5.6 | 7.8 | 9.10 | 11.12 | 13.14 |

having the *albicans* complex in common with *O. biennis*, while the *rubens* complex of *O. biennis* is

| rubens | 1.2 | 3.11 | 4.9 | 5.6 | 7.12 | 8.14 | 10.13 |

In *O. muricata* the complexes *rigens* (CATCHESIDE) and *curvans* are

| rigens | 1.2 | 3.4 | 5.6 | 7.12 | 8.14 | 9.10 | 11.13 |
| curvans | 1.14 | 2.3 | 4.6 | 7.11 | 8.9 | 10.12 | 5.13 |

Thus *rubens* of *O. biennis* and *rigens* of *O. muricata* have four pairs of chromosome ends in common. The complex *flectens* of *O. atrovirens* is

| flectens | 1.4 | 2.3 | 5.7 | 6.10 | 8.9 | 11.13 | 12.14 |

It thus has three chromosomes in common with *albicans*, two in common with *flavens*, none in common with *rigens*, but two with the same ends as in *curvans*. By comparing the chromosome ends in eastern North America with those in the Middle West it should be

---

[27] All the strains of *O. suaveolens* are heterozygous for the genes *sulfurea*, *pilosa*, *defekt* and *Maculata* (RENNER and others 1952). In one strain the *flavens* lethal factor lies in chromosome 8·9, not in 5·6.

possible to get more information regarding the sequence of segmental interchanges which have occurred. It is evident that phenotype similarity is sometimes a very poor indicator of real relationship in this genus, but determination of the sequence of segmental interchanges in different parts of the continent should throw further light on the evolution of the genus.

The extensive work of CLELAND (1950) has made progress towards the determination of lines of phylogeny in Oenothera by the study of the pairing of chromosome ends. He suggests (p. 244) that "the term 'species' can be properly applied to any group of individuals of common origin and occupying a definite geographical range, which possess a high degree of structural similarity, and which at the same time displays a recognizable distinctness from other similar groups." With such a definition, it seems clear that harmonization of the views derived from the study of chromosomal interchange and those of the advanced taxonomist will ultimately be achieved.

This work was aided by grants from the Royal Society in 1934, 1935 and 1937. The many friends who aided in the collection of seeds for these cultures have already been recorded individually in the text.

CONCLUSIONS

This survey of forty years' work with Oenothera considers first the evolution of the whole family Onagraceae. The evolutionary significance of the numerous mutations occurring in wild species of Oenothera is pointed out, also the chromosome numbers and the frequency of non-disjunction and tetraploidy. Tendencies and directions in the phylogeny of the genus Oenothera and the family Onagraceae are indicated. A taxonomic survey of Oenothera in Eastern North America, with maps of distribution, is followed by an intensive study of variation in the new species and varieties previously described. This is based upon hundreds of cultures from different localities. Cultures of the same species from many localities show numerous minor variations having a gene basis in the population but remaining generally clearly within the species. This is perhaps the most detailed study yet made of variation in any genus of flowering plants. A number of forms previously described as species by various authors are now reduced to varieties (GATES 1957).

The evolutionary significance and effect of the numerous cytological and genetical conditions originally discovered in Oenothera, or found to exist there, are considered. These include catenation (see p. 101), ring-pair chromosome formation (see p. 101), trisomics, non-disjunction, polyploid mutations, genetic complexes, balanced lethals, self-sterility factors, and self-pollination. Many of the mutations which have occurred in Oenothera in the numerous cultures of the last fifty years, especially mutations of morphological nature, are shown to have evolutionary significance.

Catenation has developed independently in *Oenothera Hookeri* on the Pacific coast, *O. grandiflora* in Alabama and *O. argillicola* in Virginia, probably beginning in the genus only in or after the Pleistocene period. Some large-flowered Mexican species also have 7 free pairs. Formation of chromosome rings has also developed independently in a number of other genera and subgenera of the Onagraceae. The diversity of species, microspecies and varieties of Oenothera, spread over North America since the retreat of the ice, appears to have arisen mainly through (1) mutations, sometimes parallel in different lines of descent (e.g., small flowers), (2) chromosome catenation, again in parallel lines of descent, and (3) occasional hybridization.

A study is made of the relative importance of measurements and non-measureable characters in taxonomic descriptions, based on large numbers of measurements in related forms. Few if any of the species or microspecies are in all localities sharp and clear-cut entities.

In the case of *O. Hazelae*, *O. ammophiloides* var. *laurensis*, *O. Victorini* and *O. deflexa* cultures from limited areas are shown to contain many minor biotypes, each of them breeding true but differing in one or more characters, thus simulating a freely interbreeding Mendelian population. Frequently forms shade into each other with transitions of one or more single-gene characters, relating them to neighbours in adjoining areas. A species in Oenothera is thus a concept or a judgment and not a rigidly fixed entity. Furthermore it can always be divided into progressively finer units by more detailed observations of more cultures from the same area. Probably the same applies to many other plant genera if an equally detailed analysis is made.

This study considers the relations between the species of the taxonomist and the microspecies and varieties or sub-species of the geneticist. It is concluded that they overlap, but both approaches are necessary for an understanding of the evolution of the genus. Even microspecies generally differ in both qualitative and quantitative characters. Further consideration of the nature of biospecies in Oenothera is not included here, especially as the nature of biospecies and paleospecies has recently been well discussed by CAIN (1956). See also SYLESTER-BRADLEY (1956). The introduction and spread of the genus in Europe is also outlined. New true-breeding forms have arisen through crossing and new mutations have also occurred.

The correlated cytological and genetical aspects lead to a general picture of the mechanism of evolution in Oenothera. It is concluded that in the pre-Pleistocene period of evolution of the genus the species had large flowers and free pairs of chromosomes. In the postglacial period catenation of chromosomes developed, associated with balanced lethals and accompanied by mutations to progressively smaller self-pollinating flowers as the genus moved northwards. Thus the maximum heterozygosity was combined with maximum seed production, and insect visitors became unnecessary. The process of catenation has developed independently several times in the evolution of different genera of the Onagraceae. On the other hand, self-sterility (S) factors have developed several times in species which have retained some free pairs of chromosomes. The method of evolution has thus been fundamentally altered between pre- and post-glacial times.

The method of chromosome pairing in the genus is acrosyndesis, the central portion of the chromosomes showing repulsion and generally remaining unpaired, their ends only showing specific pairing attractions.

In Oenothera and related genera, self-sterility, catenation and tetraploidy have all occurred repeatedly. The result has been extremely rapid diversification of the group. The early evolution of the large-flowered species preceded the beginning of catenation. Then exchange of chromosome ends began a new era in the evolution of the genus.

SUMMARY

This monograph deals with the evolution of the Onagraceae and especially of the genus Oenothera. Maps show the distribution of the many new species and varieties, chiefly in Eastern North America. Many of the mutations recorded in cultures are shown to have phylogenetic significance. Hundreds of cultures from Canada and the eastern United States show a great variety of true-breeding types whose relationships may extend in some cases north and south, in others east and west or along a coast-line, indicating lines of dispersal. An extremely intensive study of variation in *O. Hazelae* (Table III), *O. ammophiloides* (Table V), *O. Victorini* (Table VI), and *O. deflexa* (Table VII), including many cultures from limited areas, shows that each local population of these self-pollinating species contains a number of minor biotypes, thus simulating a freely interbreeding Mendelian population. Some of these differences are based on single and others on multiple genes. One must conclude that the linked chromosomes of these forms undergo exchanges of several kinds during meiosis, to produce this minor diversity.

From many hundreds of cultures of wild forms, it is concluded that the species concept of the taxonomist and the microspecies, subspecies or varieties of the geneticist overlap, the work of lumpers and splitters being complementary and not antagonistic. Innumerable minor strains can occur within a species (and sub-strains within a strain) without disturbing its integrity as a species. Even in microspecies both qualitative and quantitative characters are generally present, and the latter show increments under favourable conditions, so that the comparison of related forms has to be made under the same growing conditions i.e., side by side. With intensive observation of strains, the analysis of characters has been carried to the physiological level where the reaction of elements of the genome, e.g., the effect of temperature on petal size, can be measured.

The evolutionary conditions in Oenothera have become remarkably complex. They include non-disjunction, trisomics, polyploid mutations, catenation (linkage of the chromosomes into rings at meiosis), ring-pair chromosomes, genetic complexes, balanced lethals, self-sterility (S) factors, self-pollination and dominant mutations for smaller flowers. Several of these conditions were first discovered in Oenothera, and several, e.g. polyploidy, catenation, and S factors, have occurred independently as parallel developments in related genera.

It is concluded that the method of synapsis in Oenothera is acrosyndesis, an old term here revived to mean pairing of the chromosomes at their end portions only. The 14 chromosomes show specific attractions between their corresponding free ends. This leads to the formation of ring-pairs, which are frequently interlocked in homozygous forms, increasingly larger rings (in heterozygous species) being produced partly by exchange of non-homologous chromosome ends and partly by inter-crossing. The central portions of the chromosomes show a non-specific mutual repulsion in meiotic prophase, perhaps because they carry an excess of nucleic acid.

The large-flowered open-pollinated species in the Southern States and Mexico, with 7 free pairs of chromosomes, such as *O. Hookeri* in California, *O. grandiflora* in Alabama, and *O. argillicola* in West Virginia, are relicts dating from pre-Pleistocene times. They represent an earlier period of differentiation. The formation of chromosome rings, probably beginning in them during or after the Pleistocene, inaugurated a new era of very rapid evolution.

We may thus divide the evolution of the genus into two periods: 1. Pre-Pleistocene, probably a long period, during which differentiation of the large-flowered open-pollinated species such as *Oe. Hookeri*, *Oe. grandiflora* and *Oe. argillicola* took place by ordinary methods, the chromosomes all remaining as free pairs, 2. The Pleistocene and especially post-Pleistocene period during which chromosome linkage (catenation) took place, followed by the production of a great number of self-pollinating forms with small flowers as the genus spread northwards following the last retreat of the ice. The initiation of catenation may well have occurred during the various advances and retreats of the vegetation occasioned by the alternation of glacial and interglacial periods. These later species, having flowers which are normally self-pollinated before the flower opens, produce abundant seeds independently of insect visits and thus retain a highly heterozygous (crypthybrid) condition notwithstanding continuous self-pollination.

Thus briefly sketched, this appears to be an instance in which the method of evolution has been found to change radically in a genus during its evolution. During the earlier pre-Pleistocene period, when insect visits were necessary because of the large flowers, the long hypanthium with its nectar must have evolved in connection with the visits of moths having a long proboscis. Now (GATES 1958) bees bypass this mechanism in small-flowered species, stealing the nectar by puncturing the hypanthium at its base. They thus profit by a mechanism evolved in an earlier age but now no longer useful to the plant species. The hypanthium is thus a relic organ in the later species, harking back to a time when the biological relations of the ancestral species were quite different.

Self-sterility (S) factors have also evolved independently in several species of Oenothera and nearly related genera. They originated in the older evolution of the large-flowered species without catenation and have survived in some of the younger, small-flowered species, where their presence is now, if anything, a disadvantage to the species.

REFERENCES

ALPINUS, P., 1627. *De Plantas Exoticis.*
ARNOLD, C. G., 1955. *Z. Abst. u. Vererb.* 86: *622–662.*
ARNOLD, C. G. & BINA, H., 1957. *Flora* 144: *537–561.*
BAERECKE, M. L., 1944. *Flora* 38: *57–92.*
BARTLETT, H. H., 1913. *Rhodora,* 15: *48–53.*
BARTLETT, H. H., 1914a *Cybele Columbiana,* 1: *37–56.*
BARTLETT, H. H., 1914b. *Amer. J. Bot.,* 1: *226–243.*
BARTLETT, H. H., 1914c. *Rhodora,* 16:*33–37.*
BARTLETT, H. H., 1915a. *Bot. Gaz.,* 60: *425–456.*
BARTLETT, H. H., 1915b. *Bot. Gaz.,* 59: *81–123.*
BARTLETT, H. H., 1915c. *Amer. J. Bot.* , 2: *100–109.*
BARTLETT, H. H., 1915d., *Rhodora,* 17: *41–44.*
BELLING, J. & BLAKESLEE, A. F., 1926. *Proc. Nat. Acad. Sci.,* 12: *7–11.*
BHADURI, P. N., 1940. *Proc. Roy. Soc.,* 128B: *353–378.*
BHADURI, P. N., 1941. *Ann. Bot., N.S.,* 5: *1–14.*
BHADURI, P. N. & KAR, A. K., 1948, *Bull. Bot. Soc. Bengal,* 2: *1–14.*
BICKNELL, E. P., 1914. *Torr. Bull.,* 41: 77.
BOEDIJN, K., 1924. *Z. Abst. u. Vererb.* 32: *354–362.*
BRIDGES, C. B., 1913. *J. Exp. Zool.,* 15: *587–606.*
BRITTINGHAM, W. H., 1931. *Amer. Nat.,* 65: *121–133.*
CAIN, A. J., 1956. *Systematic Zool.,* 5: *97–109.*
CATCHESIDE, D. G., 1931. *Proc. Roy. Soc.* 109B: *165–184.*
CATCHESIDE, D. G., 1932. *Cytologia,* 4: *68–113.*
CATCHESIDE, D. G., 1933. *J. Genet.,* 27: *45–69.*
CATCHESIDE, D. G., 1935. *Genetica,* 17: *313–341.*
CATCHESIDE, D. G., 1936. *J. Genet.,* 33: *1–23.*
CATCHESIDE, D. G., 1937. *Genetica,* 19: *134–142.*
CATCHESIDE, D. G., 1940. *Proc. Roy. Soc.* 128B: *509–535.*
CATCHESIDE, D. G., 1954. *Heredity,* 8: *125–137.*
CHROMATZKA, P., 1955. *Z. Abst. u. Vererb.,* 87: *267–297.*
CLAUSEN, J., 1954. *Caryologia* Suppl. Vol: *469–479.*
CLAUSEN, J., KECK, D. D. & HIESEY, W. M., 1941. *Amer. Nat.* 75: *231–250.*
CLAUSEN, J., KECK, D. D. & HIESEY, W. M., 1948. *Carneg. Inst. Wash. Publ* No. 581, pp. *129.*
CLELAND, R. E., 1922. *Amer. J. Bot.,* 9: *391–413.*
CLELAND, R. E., 1924. *Bot. Gaz.,* 77: *149–170.*
CLELAND, R. E., 1931. *Amer. J. Bot.,* 18: *629–640.*
CLELAND, R. E., 1935. *Proc. Amer. Phil. Soc.,* 75: *339–429.*
CLELAND, R. E., 1937. *Proc. Amer. Phil. Soc.,* 77: *477–542.*
CLELAND, R. E., 1940. *Genetics,* 25: *636–644.*
CLELAND, R. E., 1944. *Amer. Nat.,* 78: *5–28.*
CLELAND, R. E., 1949. *Hereditas Suppt.,* Vol: *173–188.*
CLELAND, R. E., 1950. *Indiana Univ. Publ., Science Series,* No. 16. Pp. *348.*
CLELAND, R. E., 1951. *Evolution,* 5: *165–176.*
CLELAND, R. E., 1954. *Proc. Amer. Phil. Soc.* 98: *189–203.*
CLELAND, R. E. & BLAKESLEE, A. F., 1930. *Proc. Nat. Acad. Sci.* 16: *183–189.*
CLELAND, R. E. & BLAKESLEE, A. F., 1931. *Cytologia,* 2: *175–233.*
CLELAND, R. E. & BRITTINGHAM, W. H., 1934. *Genetics,* 19: *62–72.*
CLELAND, R. E. & OEHLKERS, F., 1930. *Jb. wiss. Bot.* 73: *1–124.*
COCKERELL, T. D. A., 1919. *Gard. Chron.* 65: *38.*
CROWE, L. K., 1955. *Heredity,* 9: *293–322.*
DARLINGTON, C. D., 1931. *J. Genet.,* 24: *405–474.*
DAVIS, B. M., 1914. *Z. Abst. u. Vererb.,* 12: *169–205.*
DAVIS, B. M., 1916. *Amer. Nat.,* 50: *688–696.*
DAVIS, B. M., 1924. *Proc. Amer. Phil. Soc.,* 63: *239–278.*

110

DAVIS, B. M., 1926. *Proc. Amer. Phil. Soc.*, 65: *349–378.*
DAVIS, B. M., 1940. *Genetics*, 25: *433–437.*
DAVIS, B. M., 1947. *Genetics*, 32: *185–199.*
DOBZHANSKY, TH., 1949. *Proc. 8th Congr. Genetics* pp. *210–224.*
EMERSON, S., 1936a. *J. Genet.*, 32: *315–342.*
EMERSON, S., 1936b. *Genetics*, 21: *200–224.*
EMERSON, S., 1938. *Genetics*, 23: *190–202.*
EMERSON, S., 1939. *Genetics*, 24: *524–537.*
EMERSON, S., 1940. *Bot. Gaz.*, 101: *890–911.*
EMERSON, S., 1941. *Genetics*, 26: *469–473.*
EMERSON, S. & STURTEVANT, A. H., 1932. *Genetics*, 17: *393–412.*
FORD, C. E., 1936. *J. Genet.*, 33: *275–303.*
GATES, R. R., 1908. *Bot. Gaz.*, 46: *1–34.*
GATES, R. R., 1909a. *Arch. Zellforsch.*, 3: *525–552.*
GATES, R. R., 1909b. *Bot. Gaz.*, 48: *179–199.*
GATES, R. R., 1910a. *Ann. Rept. Mo. Bot. Gard.*, 21: *175–184.*
GATES, R. R., 1910b. *Amer. Nat.*, 44: *203–213.*
GATES, R. R., 1911. *Z. Abst. u. Vererb.*, 4: *337–372.*
GATES, R. R., 1912a. *Nature*, 89: *659–660.*
GATES R. R. 1912b. *New Phytol.*, 11: *50–53.*
GATES, R. R., 1913. *Rhodora*, 15: *45–48.*
GATES, R. R., 1914a. *Ann. Mo. Bot. Gard.*, 1: *383–396.*
GATES, R. R., 1914b. *Z. Abst. u. Vererb.*, 11: *209–279.*
GATES, R. R., 1915. *The Mutation Factor in Evolution.* London. MacMillan.Pp.353
GATES, R. R., 1920. *Mutations and Evolution.* Cambridge Univ. Press. Pp. 118.
GATES, R. R., 1923. *J. Genet.*, 13: *13–45.*
GATES, R. R., 1923a. *Ann. Bot.*, 37: *548–563.*
GATES, R. R., 1923b. *Ann. Bot.* 37: *565–569.*
GATES, R. R., 1927. *Canad. Field-Nat.*, 41: *23–27.*
GATES, R. R., 1928. *Bibliogr. Genet.*, 4: *401–492.*
GATES, R. R., 1930. *J. Bot.*, 68: *44–46.*
GATES, R. R., 1931. The Cytological basis of mutations. *Amer. Nat.*, 65: *97–120.*
GATES, R. R., 1932. A genetic study of size inheritance. *Bull. Lab. Genetics Akad. Nauk. SSSR.*, 9: *13–28.*
GATES, R. R., 1933. *J. Linn. Soc., Bot.*, 49: *173–197.*
GATES, R. R., 1936. *Phil. Trans. Roy. Soc.*, 226B: *239–355.*
GATES, R. R., 1938. *Amer. Nat.*, 72: *340–349.*
GATES, R. R., 1939. *Nature*, 143: 245.
GATES, R. R., 1942. *Bot. Rev.*, 8: *337–409.*
GATES, R. R., 1948. *Human Ancestry.* Harvard University Press. Pp. 422.
GATES, R. R., 1950. *Canad. Field-Nat.*, 64: *142–145.*
GATES, R. R., 1951. The taxonomic units in relation to cytogenetics and gene-ecology. *Proc. VII Internat., Botan. Congress, and Amer. Nat.*, 85: *31–50.* 1951.
GATES, R. R., 1951a. *Canad. Field-Nat.*, 65: *194–197.*
GATES, R. R., 1957. *Rhodora*, 59: *9–17.*
GATES, R. R., 1958. *Rhodora*, 60:
GATES, R. R. & CATCHESIDE, D. G., 1931. *Nature*, 128: *637.*
GATES, R. R. & CATCHESIDE, D. G., 1932. *J. Genet.* 26: *143–178.*
GATES, R. R. & FORD, C. E., 1938. *Tabulae biol.*, 15: *122–153.*
GATES, R. R. & GOODWIN, K. M., 1930. *J. Genet.*, 23: *123–156.*
GATES, R. R. & GOODWIN, K. M., 1931. *Proc. Roy. Soc.* 109B: *149–164.*
GATES, R. R. & NANDI, H. K., 1935. *Phil. Trans. Roy. Soc.*, 225B: *227–254.*
GATES, R. R., & THOMAS, N., 1914. *Quart. J. micr. Sci.*, 59: *523–571.*
GECKLER, L. B., 1950. In CLELAND 1950, pp. *160–217.*
GERHARD, K., 1929. *Jena. Z. Naturw.* 64: *283–338.*
GERSHER, N., 1921. *Beih. z. Bot. Centralbl.*, 38: *204–216.*
GRANT, W. F., 1955. *Brittonia*, 8: *121–150.*

111

HAGA, T., 1944. *J. Fac. Sci. Hokkaido Imp. Univ.*, 5: *98–121.*
HAGEN, C. W., JR., 1950. In CLELAND, 1950, pp. *305–348.*
HAUSTEIN, E., 1939. *Z. Abst. u. Vererb.*, 76: *411–421.*
HAUSTEIN, E., 1952. *Z. Abst. u. Vererb.*, 84: *417–453.*
HÅKANSSON, A., 1925. *Hereditas*, 6: *257–274.*
HÅKANSSON A. 1931. *Ber. dtsch. bot. Ges.*, 49: *228–234.*
HÅKANSSON, A., 1941. *Z. Abst. u. Vererb.*, 82: *251–274.*
HÅKANSSON, A., 1941. *Hereditas*, 6: *257–274.*
HARTE, C., 1948. *Z. Abst. u. Vererb.*, 82: *495–640.*
HARTE, C. & BISSINGER, B., 1952. *ibid.*, 84: *251–269.*
HECHT, A., 1942. *Proc. Indiana Acad. Sci.*, 51: *87–93.*
HECHT, A., 1944. *Genetics*, 29: *69–74.*
HECHT, A., 1950. *Indiana Univ. Publ. Sci.* No. 16 p.p. *255–304.*
HEDAYATULLAH, S., 1932. *J. Genet.*, 26: *179–197.*
HEDAYATULLAH, S., 1933. *Proc. Roy. Soc.*, 113B: *57–70.*
HERIBERT-NILSSON, N., 1915. *Lunds Univ. Arsskrift. N.E. Avd.* 2. Vol. 12. No. 1,
    pp. *1–132.*
HOEPPENER, E. & RENNER, O., 1928. *Z. Abst. u. Vererb.* 49: *1–25.*
HOEPPENER, E. & RENNER, O., 1929. *Bot. Abhandl.* (K. Goebel) 15: *1–86.*
HUDSON, W. H., 1923. *Idle Days in Patagonia.* London.
HUNGER, F. W. T., 1913. *Ann. du Jard. de Buitenzorg*, 2nd Ser., 12: *92–113.*
JACOB, K. T., 1940. *Bot. Gaz.*, 102: *143–155.*
KISTNER, G., 1955. *Z. Abst. u. Vererb.*, 86: *521–544.*
KLEBAHN, H., 1925. *Z. Abst. u. Vererb.*, 39: *8–30.*
KÖHLER, K., 1929. *Z. Abst. u. Vererb.*, 49: *242–325.*
KULKARNI, C. G., 1929. *Bot. Gaz.*, 87: *218–259.*
LARUE, C. D. & BARTLETT, H. H., 1918. *Genetics*, 3: *207–224.*
LEWIS, D., 1947. *Heredity*, 1: *85–108.*
LEWIS, H. & LEWIS, M. E., 1955. *Univ. of Calif. Pub. Bot.*, 20: *241–392.*
LEWIS, H. & ROBERTS, M. R., 1956. *Evolution*, 10: *126–138.*
LINDER, R., 1950. *C. R. Paris*, 230: *1310–1311.*
MACDOUGAL, D. T., 1905. *Carnegie Inst. Washington. Publ.* No. 24. Pp. *57.*
MACDOUGAL, D. T., VAIL, A. M. & SHULL, G. H., 1907. *Carnegie Inst. Washington.*
    Publ. No. 81. Pp. *92.*
MARQUARDT, H., 1938. *Z. Zellforsch.*, 27: *159–210.*
MATHER, K., 1949. *Biometrical Genetics.* Dover Publications, pp. *158.*
MICHAELIS, P., 1949. *Z. Abst. u. Vererb.*, 83: *36–85.*
MICHAELIS, P., 1949. *Der Züchter*, 19: *326–331.*
M:CHAELIS, P., 1951. *Cold Spring Harbor Symposia*, 16: *121–129.*
MICHAELIS, P., 1953. *Acta Biotheoretica*, 11: *1–26.*
MICKAN, M., 1936. *Flora*, 30: *1–20.*
MUNZ, P. A., 1929. *Amer. J. Bot.*, 16: *702–715.*
MUNZ, P. A., 1931. *Amer. J. Bot.*, 18: *728–738.*
MUNZ, P. A., 1935. *Amer. J. Bot.*, 22: *645–663.*
MUNZ, P. A., 1937. *Bull. Torr. Bot. Club.*, 64: *287–306.*
MUNZ, P. A., 1940. *El Aliso*, 2: *1–47.*
OEHLKERS, F., 1921. *Z. Abst. u. Vererb.*, 26: *1–31.*
OEHLKERS, F., 1923. *Z. Abst. u. Vererb.*, 31: *201–260.*
OEHLKERS, F., 1930. *Z. Bot.*, 22: *473–537.*
OEHLKERS, F., 1935. *Z. Bot.*, 28: *161–222.*
OEHLKERS, F., 1938. *Z. Abst. u. Vererb.*, 75: *277–297.*
OEHLKERS, F. & LINNERT, G., 1949. *ibid.*, 83: *136–156.*
OELKRUG, K., 1934. *Z. Abst. u. Vererb.*, 68: *22–93.*
VAN OVEREEM C., 1922. *Beih. Bot. Centlbl.*, 39: *1–80.*
PATHAK, G. N., 1940. *Amer. J. Bot.*, 27: *117–121.*
PEASE, JOYCE, 1940. *Ann. Eugen.*, 10: *144–159.*
RENNER, O., 1917. *Z. Abst. u. Vererb.*, 18: *121–294.*

112

RENNER, O., 1921. *Ber. dtsch. bot. Ges.*, 39: *264–270*
RENNER, O., 1924. *Biol. Zbl.*, 44: *309–336.*
RENNER, O., 1925. *Biblioth. Genet.*, 9: *1–168.*
RENNER, O., 1933. *Flora*, 27: *215–250.*
RENNER, O., 1937a. *Z. Abst. u. Vererb.*, 74: *91–124.*
RENNER, O., 1937b. *Cytologia* (Fujii Vol.): *644–653.*
RENNER, O., 1937c. *Flora*, 31: *182–226.*
RENNER, O., 1939. *Flora*, 33: *215–238.*
RENNER, O., 1941. *Flora*, 35: *201–238.*
RENNER, O., 1942a. *Flora*, 36: *117–214.*
RENNER, O., 1942b. *Ber. dtsch. bot. Ges.*, 60: *448–466.*
RENNER, O., 1943a. *Z. Bot.*, 39: *49–105.*
RENNER, O., 1943b. *Flora*, 36: *325–334.*
RENNER, O., 1943c. *Z. Abst. u. Vererb.*, 81: *391–483.*
RENNER, O., 1943d. *Flora*, 37: *216–229.*
RENNER, O., 1945. *Arch. J. Klaus-Stift.* 20 Suppt.: *164–184.*
RENNER, O., 1949. *Z. Abst. u. Vererb.*, 83: *1–25.*
RENNER, O., 1951. *Ber. dtsch. bot. Ges.*, 63: *129–138.*
RENNER, O. & CLELAND, R. E., 1934. *Z. Abst. u. Vererb.*, 66: *275–318.*
RENNER, O. and others, 1952. *Z. Naturforsch.* 7b: *368–371.*
RENNER, O. & SENSENHAUER, R., 1942. *Z. Abst. u. Vererb.*, 80: *570–589.*
RHOADES, M. M., 1933. *J. Genet.*, 27: *71–93.*
RUDLOFF, K. F., 1929. *Z. Abst. u. Vererb.*, 52: *191–235.*
SCHWEMMLE, J., 1926. *Jb. wiss. Bot.*, 65: *778–818.*
SCHWEMMLE, J., 1927. *Jb. wiss. Bot.*, 66: *579–595.*
SCHWEMMLE, J., 1928a. *Jb. wiss. Bot.*, 67: *849–876.*
SCHWEMMLE, J., 1928b. *Ber. dtsch. bot. Ges.*, 46: *552–559.*
SCHWEMMLE, J. & ZINTL, M., 1939. *Z. Abst. u. Vererb.*, 76: *353–410.*
SHEFFIELD, F. M. L., 1927. *Ann. Bot.*, 41: *779–816.*
SHEFFIELD, F. M. L., 1929. *Proc. Roy. Soc.*, 105B: *207–230.*
SHULL, G. H., 1921. *J. Hered.*, 12: *354–363.*
SHULL, G. H., 1926. *Genetics*, 11: *201–234.*
SHULL, G. H., 1927. *Proc. Nat. Acad. Sci.*, 13: *21–24.*
SHULL, G. H., 1928a. *Z. Abst. u. Vererb.*, 46: *35.*
SHULL, G. H., 1928b. *Proc. Nat. Acad. Sci.*, 14: *142–146.*
SHULL, G. H., 1932. *Z. Abst. u. Vererb.*, 60: *219–234.*
SHULL, G. H., 1934. *Amer. Nat.*, 68: *481–490.*
SHULL, G. H., 1937. *Amer. Nat.*, 71: *69–82.*
SIKKA, S. M., 1940. *J. Genet.*, 39: *809–834.*
SLOATMAN, R. J., 1953. *Amer. J. Bot.*, 40: *835–836.*
SMITH, J. D. & ROSE, J. N., 1913. *Contrib. U.S.Nat. Herb.*, 16: *287–298.*
STEINER, E., 1951. *Evolution*, 5: *265–272.*
STEINER, E., 1952. *Evolution*, 6:*69–80.*
STEINER, E. & SCHULTZ, M. S., 1958. *Science* **127**: *516–517.*
STINSON, H. T., 1953. *Genetics*, 38: *389–406.*
STINSON, H. T. & STEINER, E., 1955. *Amer. J. Bot.*, 42: *905–911.*
STOMPS, TH., 1913. *Ber. dtsch. bot. Ges.*, 31: *166–172.*
STOMPS, TH., 1957. *Acta Botan. Neerlandica* **6**: *378–380.*
STURTEVANT, A. H., 1931. *Z. Abst. u. Vererb.*, 59: *365–380.*
SYLVESTER-BRADLEY, P. C., (Ed.)., 1956. *The species concept in Palaeontology.*
London, Systematics Assoc. pp. 145.
TANDON, S. L., 1956. Cytogenetic evidence for the inclusion of *Oenothera affinis*
Camb. under *O. mollissima* L. *Cytologia* **21**: *252–271.*
TANDON, S. L. & HECHT, A., 1953. *Cytologia*, 18: *133–145.*
TURESSON, G., 1925. *Hereditas*, 6: *147–236.*
VAIL, A. M., 1905. *Torreya*, 5: *9–10.*
VASEK, F. C., 1956. *Amer. J. Bot.*, 43: *366–371.*

VAVILOV, N. I., 1922. *J. Genet.*, 12: *47–89.*
DE VRIES, HUGO, 1918. *Genetics*, 3: *1–26.*
DE VRIES, HUGO, 1925. *Genetics*, 10: *211–222.*
DE VRIES, HUGO, 1929. *Z. Abst. u. Vererb.*, 52: *121–190.*
WARTH, G., 1925. *Z. Abst. u. Vererb.*, 38: *200–257.*
WEIDNER-RAUH, E., 1939. *Z. Abst. u. Vererb.*, 76: *422–486.*
WEIN, K., 1936. *Beih. z. Bot. Centralbl.* 55: *419-543.*
WISNIEWSKA, E., 1935. *Acta Soc. Bot. Pol.*, 12: *113–164.*

INDEX